Effect of IL-10 and anti-TGF-beta
antibodies on the morphology of
bone marrow stroma cultures

from
Interleukin-10
by
Jan E. DeVries and
Rene de Waal Malefyt
© R.G. Landes Co. 1995

MOLECULAR
BIOLOGY
INTELLIGENCE
UNIT

MOLECULAR BIOLOGY AND PHARMACOLOGY OF THE ENDOTHELINS

Gillian A. Gray, Ph.D.
David J. Webb, M.D.
University Department of Medicine
Western General Hospital
Edinburgh, Scotland, U.K.

Springer
New York Berlin Heidelberg London Paris
Tokyo Hong Kong Barcelona Budapest

R.G. LANDES COMPANY
AUSTIN

MOLECULAR BIOLOGY INTELLIGENCE UNIT

MOLECULAR BIOLOGY AND PHARMACOLOGY OF THE ENDOTHELINS

R.G. LANDES COMPANY
Austin, Texas, U.S.A.

Submitted: March 1995
Published: July 1995

Please address all inquiries to the Publisher:
R.G. Landes Company, 909 Pine Street, Georgetown, Texas, U.S.A. 78626
or
P.O. Box 4858, Austin, Texas, U.S.A. 78765
Phone: 512/ 863 7762; FAX: 512/ 863 0081

U.S. and Canada ISBN 1-57059-261-6

International ISBN 3-540-59425-6

PUBLISHER'S NOTE

R.G. Landes Company publishes five book series: *Medical Intelligence Unit, Molecular Biology Intelligence Unit, Neuroscience Intelligence Unit, Tissue Engineering Intelligence Unit* and *Biotechnology Intelligence Unit.* The authors of our books are acknowledged leaders in their fields and the topics are unique. Almost without exception, no other similar books exist on these topics.

Our goal is to publish books in important and rapidly changing areas of medicine for sophisticated researchers and clinicians. To achieve this goal, we have accelerated our publishing program to conform to the fast pace in which information grows in biomedical science. Most of our books are published within 90 to 120 days of receipt of the manuscript. We would like to thank our readers for their continuing interest and welcome any comments or suggestions they may have for future books.

Deborah Muir Molsberry
Publications Director
R.G. Landes Company

CONTENTS

SECTION I: MOLECULAR BIOLOGY

SECTION II: PHARMACOLOGY

ABBREVIATIONS

ACE	angiotensin converting enzyme
ANG II	angiotensin II
ANP	atrial natriuretic peptide
big ET	big endothelin
bp	base-pair
Ca^{2+}	calcium ion
cAMP	cyclic adenosine monophosphate
cDNA	complementary DNA
CBF	cerebral blood flow
CHF	congestive heart failure
CHO	chinese hamster ovary
DAG	diacylglycerol
DNA	deoxyribonucleic acid
DOCA	deoxycorticosterone acetate
ECE	endothelin-converting enzyme
EHT	essential hypertension
ET	endothelin
ET_A	endothelin-A type receptor
ET_B	endothelin-B type receptor
cGMP	cyclic guanosine monophosphate
HUVEC	human umbilical vein endothelial cells
IP_3	inositol triphosphate

kD	kilodaltons
MAPK	mitogen-activated protein kinase
NO	nitric oxide
NEP	neutral endopeptidase
PA	phosphatidic acid
PGE_2	prostaglandin E_2
PGI_2	prostacyclin
PKC	protein kinase C
PLA_2	phospholipase A_2
PLC	phospholipase C
PLD	phospholipase D
PTK	protein tyrosine kinase
RNA	ribonucleic acid
ROC	receptor operated channel
SAH	sub-arachnoid hemorrhage
SHR	spontaneously hypertensive rats
SRTX	sarafotoxin
TXA_2	thromboxane A_2
VOC	voltage operated channel
VSMC	vascular smooth muscle cells
WKY	Wistar-Kyoto rats

in scope to that associated with the angiotensin converting enzyme inhibitors and calcium entry blockers, can be envisaged.

In writing this book, we are very aware that there is much about the ET system that remains to be understood. However, the basic molecular biology and pharmacology of the ET system is now well established and there are clear indications that the ETs are involved in both the physiological regulation of the cardiovascular system and in the pathophysiology of cardiovascular disease. This book provides a current update of the molecular biology and pharmacology of the ETs and indicates the therapeutic areas for which the development of inhibitors of ET production, or antagonists at ET receptors, are likely to prove fruitful.

ACKNOWLEDGMENT

The authors wish to thank the British Heart Foundation, British Pharmacological Society and the Scottish Office for supporting their research on the endothelins. Appreciation is also extended to Emma Mickley, Paula Smith and the Medical Illustration staff of the Department of Medicine, Edinburgh University for their assistance in the preparation of figures.

MOLECULAR BIOLOGY

CHAPTER 1

DISCOVERY OF THE ENDOTHELIN FAMILY OF PEPTIDES

David J. Webb

The past 10 years have seen a rapid increase in our understanding of the cardiovascular system and provided additional insights into the subtlety and complexity of the interactions between the many control systems. Recently, in particular, the endothelial cells have become a prominent target for research. This monolayer of cells lining the blood vessels is now known to play a central role in regulation of coagulation, lipid transport, immunological reactivity and vascular tone. In recent years, one of the most important emerging principles in vascular biology has been the fundamental role played by the endothelium as both a sensor and effector of local vascular tone. This chapter provides the background leading up to the discovery of the potent vasoconstrictor and pressor peptide, endothelin (ET), outlines the cardiovascular actions of the ETs, and summarizes the key developments in ET research.

HISTORICAL BACKGROUND

Over the past 20 years it has become clear that many key factors regulating vascular tone are produced by the endothelial cells lining all blood vessels (Fig. 1.1). Soon after the discovery of prostacyclin,[1] Furchgott and Zawadski[2] demonstrated that the vascular relaxation induced by acetylcholine is dependent on the presence of the endothelium and mediated by a labile nonprostaglandin humoral factor, later known as endothelium-derived relaxing factor (EDRF), and subsequently identified as nitric oxide.[3,4]

Molecular Biology and Pharmacology of the Endothelins edited by Gillian A. Gray and David J. Webb. © 1995 R.G. Landes Company.

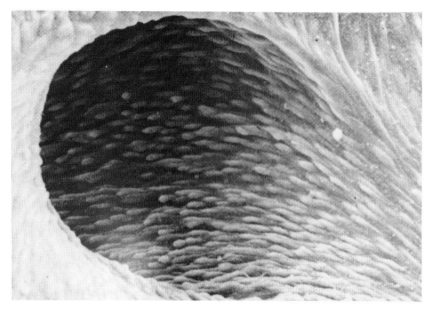

Fig. 1.1. Scanning electron photomicrograph of a human blood vessel illustrating the endothelial cell monolayer lining the inner surface. Reproduced with permission from Haynes et al, Brit J Hosp Med 1992; 47: 340-9.

The isolation of EDRFs led to a search for counterbalancing endothelium-derived constricting factors or EDCFs. In 1984, O'Brien and McMurtry[5] reported that pulmonary arterial and aortic endothelial cells secrete a polypeptide causing sustained contraction of bovine pulmonary arteries and in the following year Hickey and colleagues[6] reported production of a similarly sustained EDCF from bovine aortic endothelial cells in culture. The substance, derived from conditioned medium, caused a dose-dependent contraction of porcine, bovine and canine coronary arteries, apparently dependent on extracellular calcium. This substance was shown to be insensitive to alpha- and beta-adrenergic, serotonergic, histaminergic and cholinergic receptor antagonists, and inhibitors of cyclo-oxygenase and lipoxygenase pathways. Being protease sensitive, it also had the characteristics of a polypeptide. By 1987, this EDCF had been further characterized, confirming its structure as a peptide with molecular weight ~3000 daltons.[7] These experiments demonstrated additional insensitivity to opioid, leukotriene, angiotensin II and substance P receptor antagonists, suggesting a previously unknown mediator was involved, consistent with its uniquely sustained action. Vasoconstriction could be partially reversed by the calcium antagonist, verapamil, and completely abolished by isoprenaline and sodium nitroprusside.

THE DISCOVERY OF THE ENDOTHELINS

In their landmark paper, published in 1988, Yanagisawa and colleagues[8] at Tsukuba University, Japan, reported an elegant series of experiments in which they isolated and characterized an extremely potent vasoconstrictor peptide from the culture supernatant of porcine aortic endothelial cells, naming it *endothelin* (Fig. 1.2).

After initial purification from conditioned culture medium, based on a porcine coronary artery bioassay, the amino acid composition of ET, now known as ET-1, was determined by acid hydrolysis, and the sequence elucidated using automated gas-phase peptide sequencing and carboxyterminal analysis by hydrazinolysis. This gave a relative molecular mass (M_r) of 2,492, and a composition of 21 amino acid residues with free amino- and carboxy-terminals. The four cysteine residues were found to form two intra-chain disulphide bridges linking amino cysteine residues 1 to 15 and 3 to 11 (Fig. 1.2). Synthetic ET-1, based on this structure, showed identical retention times on HPLC to the natural peptide and full biological activity.

ET-1 was shown to act as a potent vasoconstrictor in blood vessels from a wide range of species and a number of different anatomical locations, including arteries and veins. The EC_{50} for ET-1 in porcine coronary artery strips, at 4×10^{-10} M, was at least an order of magnitude lower than for angiotensin II, vasopressin or neuropeptide Y. ET-1 remains the most potent mammalian vasoconstrictor yet identified. Consistent with the earlier observations concerning the potent EDCF present in endothelial cell culture supernatants, ET-1 caused a slowly developing and sustained vasoconstriction which proved difficult to wash out but could be reversed completely by isoprenaline or glyceryl trinitrate. Also, vasoconstriction was resistant to antagonists at α-adrenergic, H_1-histaminergic and serotonergic receptors and inhibitors of cyclo-oxygenase and lipoxygenase, consistent with a novel action on vascular smooth muscle. Contraction was completely inhibited in Ca^{2+}-free medium containing 1 mM EGTA and markedly attenuated in the presence of the dihydropyridine Ca^{2+}-channel blocking agent, nicardipine, suggesting that influx of extracellular calcium was important for its action. In vivo, intravenous bolus injections of ET-1 caused a sustained rise in arterial pressure in anesthetized, chemically denervated rats, typically requiring more than 40-60 minutes to return to baseline levels (see Fig. 6.1). An initial transient depressor effect of ET-1 was apparent from the original study but did not receive comment. It is now known that release of endothelium-dependent vasodilators, including nitric oxide, prostacyclin and the endothelium-derived hyperpolarizing factor account for this effect (Fig. 1.3: see chapter 6). Taken together with their identification and sequencing of the ET-1 gene, this work[8] represented a remarkable *tour de force* for molecular biology and, along with the pharmacological observations, opened the way for a major new field of endothelial cell biology.

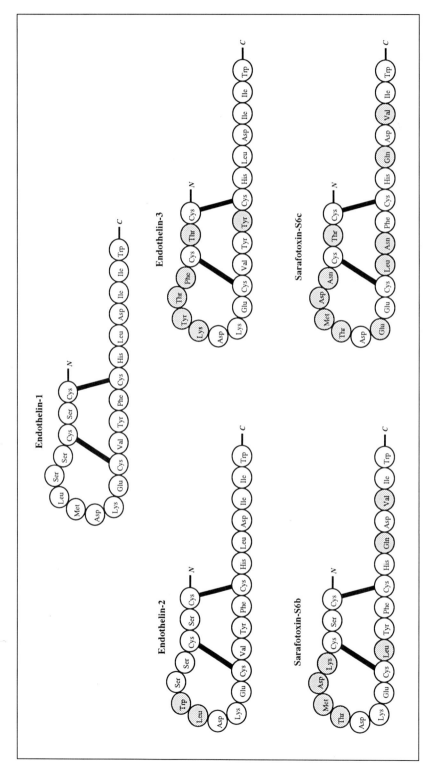

Fig. 1.2. Structure he ET-1, ET-2 and ET-3, and of sarafotoxins S6b and S6c.

In order to find the gene for ET-1, a porcine aortic endothelial cell cDNA library was screened using an 'optimal' synthetic probe encoding residues 7-20 of ET-1. Hybridizing clones were selected for determination of the complete nucleotide structure of the gene, encoding a 203 amino acid prepro-form of the mature peptide (see Fig. 2.1). The first 19 residues of the deduced amino acid sequence of prepro ET-1 are characteristic of a secretory signal sequence, creating a 184 amino acid proET-1 peptide. Paired basic amino acid residues Lys 51-Arg 52 and Arg 92-Arg 93, which are recognized by processing endopeptidases, encompass not only the ET sequence but also an additional amino acid sequence, so that ET-1 is not generated directly but produced in the form of a so-called 'big ET-1.' This is then cleaved at Trp 73-Val 74 to form the mature peptide. This unusual proteolytic processing, by one or more 'ET converting enzymes' (ECEs), may prove an important target for anti-ET therapy.

Control of ET-1 production, and the capacity for its induction, was also examined. Northern blot analysis of mRNA extracted from porcine endothelial cells grown as confluent monolayers in culture showed that prepro ET-1 mRNA increases, within one hour, in response to exposure to thrombin, adrenaline and the Ca^{2+} ionophore, A23187. Given that vascular endothelial cells produce mRNA encoding a prepro-form of ET-1 have few secretory granules when examined under electron microscopy,[9] and have little vasoconstrictor activity on lysis,[7] Yanagisawa and colleagues[8] thought it unlikely that ET-1 is stored in secretory granules and released in response to stimuli. Although it seems that production is likely to be regulated predominantly at the level of transcription, some of the more recent evidence conflicts with this view (see chapter 2) and the matter presently remains unresolved.

These wide-ranging observations, covering the pharmacology and molecular biology of ET-1, have been confirmed and extended by subsequent workers. In addition to their immediate observations, Yanagisawa and colleagues also made some tentative proposals about the physiological implications of their work.[8] Noting the similarity of the structure of ET-1 to a range of venoms and toxins including scorpion toxins that can bind to tetrodotoxin-sensitive Na^+ channels, the dependence of vasoconstriction on external Ca^{2+}, and the sensitivity of vasoconstriction to nicardipine, they speculated that ET-1 might be an endogenous agonist of dihydropyridine-sensitive calcium channels. Although it has not been possible to sustain this hypothesis (see chapters 4 and 5), ETs have since been shown to have very close structural and functional similarity to sarafotoxin venom peptides produced by the Israeli burrowing asp, *Actractaspis engaddensis*[10,11] (Fig. 1.2) and these have subsequently proved particularly useful as pharmacological tools for identifying ET receptor subtypes, and their actions, in different tissues. Yanagisawa and co-workers also speculated that ET-1 might play an important role in regulation of arterial pressure and/or local blood

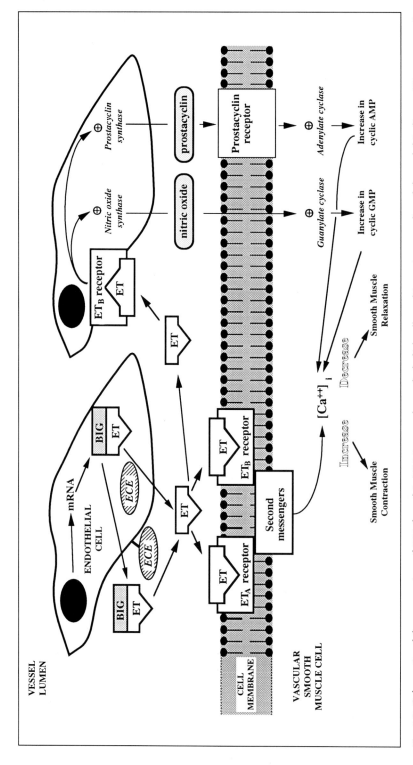

Fig. 1.3. Schematic of the generation of ET-1 from big ET-1 in vascular endothelial cells and its actions mediated through binding to ET_A receptors in vascular smooth muscle and ET_B receptors in both endothelium and smooth muscle.

flow, and that disturbances of the control of ET production could contribute to the pathogenesis of hypertension and vasospasm. More recent work has confirmed a major role for ET-1 in the maintenance of basal vascular tone and blood pressure (see chapter 6), and studies in both animal models of cardiovascular diseases and in patients serve to sustain the view that ET-1 will be an important target for the treatment of cardiovascular disease (see chapter 7).

THE ENDOTHELIN FAMILY OF PEPTIDES

The peptide that Yanagisawa and colleagues originally isolated, ET-1, is now known to be a member of a family of three ET isopeptides, ET-1, ET-2 and ET-3, each a 21 amino acid polypeptide. They are each represented as a single copy gene on separate chromosomes in the human genome (see chapter 2). ET-1, like the other members of the family, has a hydrophobic carboxy terminus and a globular amino terminus which is cross-linked by two disulphide bridges, between residues 1 and 15, and 3 and 11, respectively. Its crystal structure has recently been described (Fig. 1.4).[12] ET-1 and ET-2 differ by only 2 amino acids, at positions 6 and 7. ET-3 differs by 6 amino acids, at positions 2, 3-7, and 14, accounting for the important differences in affinity for different receptor subtypes between ET-1 and ET-3 (Fig. 1.2).

RELEVANCE AND ACTIONS

It is now known that ET-1 is the major ET isoform produced by vascular endothelial cells and probably the most important in the cardiovascular system. A large number of factors have been identified that increase ET-1 production; there are only a small number recognized to inhibit production. One isoform of ECE, so-called ECE-1 has now been cloned and characterized[13] and appears to be involved in intracellular processing of big ET-1 (Fig. 1.3). ET-1 acts on at least two receptors (see chapters 3 and 4), both of which have been cloned in humans,[14,15] and which are widely distributed in the cardiovascular system. These receptor subtypes are known, on the basis of their isoform selectivity, as the ET_A (ET-1 > ET-2 > ET-3) and ET_B receptors (ET-1 = ET-2 = ET-3). ET_A receptors are clearly of importance in generating smooth muscle contraction, but in some vessels it is now clear that ET_B receptors can also serve this purpose (Fig. 1.3; see chapter 6). ET_B receptors are also found on endothelial cells, where they act to stimulate release of endothelium-dependent vasodilator substances, and so modulate the vasoconstrictor actions of ET-1. ET-1 is primarily a paracrine and autocrine substance, and only rarely acts as a circulating hormone. ET-1 has vasoconstrictor, venoconstrictor, inotropic and pressor actions of a characteristically sustained nature. ET-1 also has vascular and cardiac growth promoting properties.

Fig. 1.4. Three dimensional structure of ET-1 derived from X-ray crystallography.

(A) Ribbon drawing showing folding of the endothelin-1 polypeptide chain. The N-terminal residues form an extended beta strand, whereas the C-terminal residues form a long, somewhat irregular, helix. The central region contains a hydrogen-bonded loop between residues 7 and 11. The N- and C-termini are linked by disulphide bonds, between residues 1 and 15, and 3 and 11, respectively.

(B) Space filling model of endothelin-1 which has been color-coded to show conserved variable residues. Key: blue = fully conserved in all the endothelins, yellow = conservatively replaced residues (Ser 2, Ser 5, Phe 14), red = variable residues (Ser 4, Leu 6, Met 7). The model is in a comparable orientation to the ribbon drawing. Reproduced with permission from Janes et al, Nature Structural Biol 1994; 1: 311-319.

FURTHER DEVELOPMENTS

The actions of ET-1, and its sites of production, have led to considerable speculation over the past six years that ET-1 may be important in the maintenance of vascular tone and blood pressure. These views are now strongly supported on the basis of recent research (see chapter 6). It has also been suggested that ET-1 may be of importance in a range of cardiovascular diseases associated either with sustained vasoconstriction, such as hypertension and heart failure, or with vasospasm, such as in diseases as disparate as subarachnoid hemorrhage, unstable and variant angina, acute renal failure and Raynaud's disease. There is a now substantial body of evidence showing that ET-1 contributes to these diseases. In addition, there is considerable evidence in animals, and a growing body of evidence in patients with cardiovascular disease, that anti-endothelin therapy, in the form of ECE inhibitors or endothelin receptor antagonists, can reverse vasospasm and ameliorate the tissue damage associated with many of these diseases (see chapter 7).

Since 1988, ET-1 has generated a large body of academic research, much of which indicates the potential for anti-endothelin therapies in cardiovascular disease. The pharmaceutical industry has contributed substantially to this research, not least by the development of effective and specific antagonists at endothelin receptors. Many of these drugs have been used to increase our understanding of the role of ET-1 in the cardiovascular system and, more recently, a range of orally active agents have been described that have the potential to be developed clinically as novel treatments for cardiovascular disease. This book provides a detailed analysis of the molecular biology and pharmacology of the endothelins in the cardiovascular system, and highlights the recent developments in understanding of the role of ET-1 in cardiovascular physiology and pathophysiology.

REFERENCES

1. Moncada S, Gryglewski R, Bunting S, Vane JR. An enzyme isolated from arteries transforms prostaglandin endoperoxides to an unstable substance that transforms platelet aggregation. Nature 1976; 263: 663-665.
2. Furchgott RF, Zawadzki JV. The obligatory role of endothelial cells in the relaxation of arterial smooth muscle by acetylcholine. Nature 1980; 288: 373-376.
3. Palmer RM, Ferrige AG, Moncada S. Nitric oxide release accounts for the biological activity of endothelium-derived relaxing factor. Nature 1987; 327: 524-526.
4. Palmer RM Ashton DS, Moncada S. Vascular endothelial cells synthesize nitric oxide from L-arginine. Nature 1988; 333: 664-666.
5. O'Brien RF, McMurtry IF. Endothelial cell supernatants contract bovine pulmonary artery rings. Am Rev Resp Dis 1984; 129: A337.

6. Hickey KA, Rubanyi G, Paul RJ, Highsmith RF. Characterisation of a coronary vasoconstrictor produced by cultured endothelial cells. Am J Physiol 1985; 248: C550-C556.

7. O'Brien RF, Robbins RJ, McMurtry IF. Endothelial cells in culture produce a vasoconstrictor substance. J Cell Physiol 1987; 132: 263-270.

8. Yanagisawa M, Kurihawa H, Kimura S, et al. A novel potent vasoconstrictor peptide produced by vascular endothelial cells. Nature 1988; 332: 411-415.

9. Johannessen JV. Electron Microscopy in Human Medicine Volume 5, McGraw-Hill, New York 1980: pp. 87-154.

10. Kloog Y, Sokolovsky M. Similarities in mode and sites of action of sarafotoxins and endothelins. Trends Pharmacol Sci 1989; 10: 212-214.

11. Lee CY, Chiappinelli VA. Sequence homology between sarafotoxins S6 and porcine endothelin (letter). Toxicon 1989; 27: 277-279.

12. Janes RW, Peapus DH, Wallace BA. The crystal structure of human endothelin. Nature Structural Biol 1994; 1: 311-319.

13. Xu D, Emoto N, Giaid A, et al. ECE-1: a membrane bound metalloprotease that catalyzes the proteolytic activation of big endothelin-1. Cell 1994; 78: 473-485.

14. Arai H, Hori S, Aramori I, et al. Cloning and expression of a cDNA encoding an endothelin receptor. Nature 1990; 348: 730-732.

15. Sakurai T, Yanagisawa M, Takuwa Y, et al. Cloning of a cDNA encoding a non-isopeptide selective subtype of the endothelin receptor. Nature 1990; 348: 732-735.

=CHAPTER 2 =

GENERATION OF ENDOTHELIN

Gillian A. Gray

In their initial description of the constricting factor released into the supernatant of cultured porcine aortic endothelial cells, Yanagisawa and colleagues[1] described isolation of the 21 amino acid peptide, endothelin (ET). Construction of a cDNA library from porcine endothelial cells and screening with an oligonucleotide probe permitted isolation, cloning and sequencing of the cDNA for the porcine peptide. The deduced sequence of a 203 amino acid precursor pro-peptide made clear the requirement for a two-stage proteolytic process to generate the mature ET peptide (Fig. 2.1). While initial processing by well characterized dibasic pair endopeptidases could account for reduction of the pro-ET to a 39 amino acid pro-peptide, termed 'big-ET', final generation of ET seemed to require a novel proteolytic process. This chapter will review the current understanding of ET generation, from characterization of the ET genes and regulation of their transcription to the characterization and recent cloning of a putative 'endothelin-converting enzyme'.

STRUCTURAL ORGANIZATION AND
TRANSCRIPTIONAL REGULATION OF THE ET GENES

The cDNA for human prepro ET was first isolated from a placental cDNA library.[2] Although the human and porcine cDNAs had only 79% sequence homology, proteolytic processing of the encoded peptides was found to result in formation of a common 21 amino acid mature peptide.[2,3] Following the sequencing of both human and porcine cDNAs, a prepro ET related gene from rat was cloned and sequenced, the predicted 21 amino acid sequence of which differed by 6 residues from human and porcine mature ET.[4] Southern blot analysis of human, porcine and rat genomic DNA subsequently provided evidence for the existence of at least three genes coding for 'ET-like'

Molecular Biology and Pharmacology of the Endothelins edited by Gillian A. Gray and David J. Webb. © 1995 R.G. Landes Company.

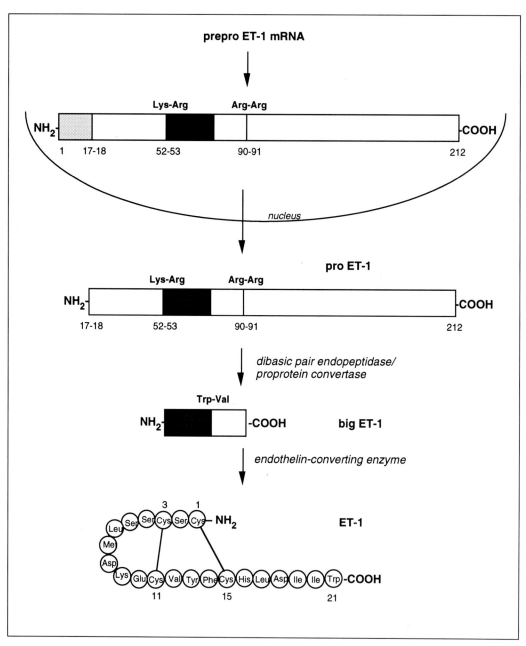

Fig. 2.1. Schematic of the proteolytic processing pathway for human endothelin-1 (ET-1). The amino terminal signal peptide sequence (amino acids 1-17, pale hatching) of the 212 amino acid prepro ET-1 is cleaved on secretion of the peptide from the nucleus. The 38 amino acid big ET-1 is then formed by proteolytic cleavage of prepro ET-1 at paired dibasic residues by proprotein convertase enzymes. Mature 21 amino acid ET-1 (dark hatching) is formed from big ET-1 through cleavage at Trp21-Val22 by endothelin-converting enzyme. Adapted from Yanagisawa et al, 1988.[1]

sequences in mammalian genomes.[5] The amino acid structure deduced from the third, previously undescribed, gene was predicted to differ by two amino acids from that of 'porcine' and 'human' ET. These three genes code for the now recognized members of the ET family of peptides (Fig. 2.2). ET-1 corresponds to the originally described porcine and human peptides, ET-3 to the rat peptide and ET-2, to the novel gene identified in the genomic DNA. All three genes are encoded in the human genome,[5] the ET-1 gene on chromosome 6,[3,6] the ET-2 gene on chromosome 1[7] and the ET-3 gene on chromosome 20.[8] Cloning and sequence analysis of the mouse genome suggested the existence of a fourth ET-like gene which was expressed exclusively in the intestine.[9] The peptide deduced from this sequence was named vasoactive intestinal constricting peptide (VIC). The same gene sequence was also found in rat tissues[9] and is now believed to be an isoform of the ET-2 gene.[7]

Human cDNA for prepro ET-1 has a 636 (bp) base pair coding region corresponding to 212 amino acids with a 5' untranslated flanking region approximately 250 base pairs long and a 3' untranslated region approximately 1127 base pairs long.[3] Prepro ET-2 and prepro ET-3 are encoded in 534 and 714 bp respectively.[7,8]

REGULATORY REGIONS OF THE ET GENES

The majority of research into regulation of ET synthesis has focused on the ET-1 gene. Therefore it should be noted at this point that, unless specified, the discussion refers to regulation of ET-1 synthesis.

All eukaryotic genes contain sequences involved in regulation of gene transcription which are important in directing RNA polymerase II binding. These regulatory regions govern basal expression of the gene and are the means by which external factors are able to modulate transcription. The ET-1 gene promoter region contains several elements commonly found in eukaryotic promoters, including a 'TATA box' and a 'CAAT' element located 31 bp and 86 bp, respectively, upstream of the transcription start site.[10] The presence of two other regions of the promoter necessary for high level transcription of the ET-1 gene was revealed by transfection of constructs containing varying prepro ET-1 5' flanking sequences into cultured bovine aortic endothelial cells.[11,12] The first of these two *cis*-nucleotide sequences, located between 148 and 117 bp upstream of the transcription start site, is believed to be a member of the family of nuclear proteins binding to the GATA motif. Transient overexpression of the GATA-2 binding protein in bovine aortic endothelial cells and HeLa cells increases the expression of ET-1.[13,14] In contrast, deletion of the GATA element from the prepro ET-1 gene in COS cells increases levels of reporter gene expression,[15] suggesting that prepro ET-1 expression might be controlled by different mechanisms in different cells. The second

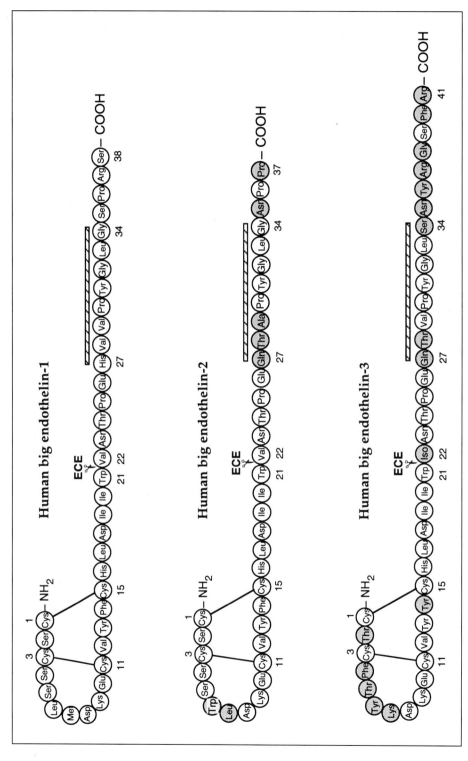

Fig. 2.2. Amino acid sequences of the three precursor (big ET) forms of human endothelin. Cysteine residues that form disulfide bonds are linked by cross bridges. Shaded circles represent amino acids that differ from the big ET-1 isoform. ECE ✂ represents the cleavage site of the precursor peptides by endothelin-converting enzyme. The shaded bar signifies the amino acids (27-34) proposed to be most important for selective conversion by the bovine endothelial ECE.[96]

promoter region is located between 117 and 98 bp upstream of the transcription start site. Selective nucleotide removal in this part of the promoter using site-directed mutagenesis suggests that an 8-bp DNA sequence (GTGACTAA, bp 109 to 102) is necessary for basal transcription of the prepro ET-1 gene in endothelial cells in culture.[16] This sequence is identical to the well characterized AP-1 protein binding site, which mediates hormone and growth factor responsiveness in association with other eukaryotic genes (reviewed in ref. 17). Using anti-Fos and anti-Jun antisera, Lee et al[16] have demonstrated that the Fos and Jun nuclear protein binding families bind to the ET-1 AP-1 sequence and that ET expression is stimulated via this sequence by c-*fos* and c-*jun*.

GATA and AP-1 sites are commonly found together with erythroid-specific genes and have been implicated in cell type specific expression of these genes.[18] However, the GATA binding protein found in endothelial cells is expressed in a wide variety of other cells[19] suggesting that it cannot be primarily responsible for the restricted pattern of expression of the ET-1 gene. This pattern could, however, be conferred by binding of a tissue specific factor to the AP-1 site, as is the case for regulation of erythroid specific β-globin HSII region.[18]

Other sequences of interest in the noncoding region of the human prepro ET gene are the 20 base pair nuclear factor 1 (NF-1) binding sequence, which may mediate the induction of prepro ET-1 by transforming growth factor β, and the hexanucleotide sequences for the acute phase reactant regulatory elements, which may be involved in the induction of ET-1 under acute physiological stress in vivo.[3,20]

Benatti et al[15] recently isolated two forms of prepro ET-1 cDNA from a human placental cDNA library. The second cDNA sequence has an 80 nucleotide extension at the 5' end, containing additional sites for initiation of gene transcription and a second promoter region upstream of the previously mapped region. The authors suggest that the two promoter regions control the transcription of two forms of prepro ET-1 mRNA.

FACTORS INFLUENCING TRANSCRIPTION

ET-1 production, detected either by measurement of mRNA or elevated immunoreactive ET concentrations, is regulated by a number of factors, some stimulatory and some inhibitory (Fig. 2.3). Stimulation of ET formation by **thrombin** occurs both in vitro, in cultured cells,[21-24] and in vivo[25] and may have a role in thrombosis.[26] Thrombin elicits cellular responses via activation of phospholipase C leading to stimulation of inositol phosphate hydrolysis, liberation of intracellular calcium and activation of protein kinase C (PKC). All of these processes may play a role in regulation of ET synthesis since thrombin stimulated liberation of ET-1 by cultured endothelial cells is prevented by inhibition of phospholipase C or calcium chelation,[21] and

by PKC inhibitors or down-regulation of PKC.[21-24] Similar mechanisms
seem to be involved in stimulation of prepro ET-1 mRNA by low
density lipoprotein, angiotensin II, vasopressin[27-30] and by ET-1 itself.[31,32]
Direct activation of PKC by **phorbol ester** results in enhancement of
ET release by porcine aortic endothelial cells.[22,24] The proto-oncogenes
c-*jun* and c-*fos* have been shown to mediate responsiveness to agents
which function through the PKC pathway.[17] Since c-*jun* and c-*fos* bind
to the AP-1 transcription regulatory element of the ET-1 promoter,[16]
it seems likely that they mediate the stimulation of ET-1 synthesis by
agents which activate PKC. The observation that thrombin stimulates

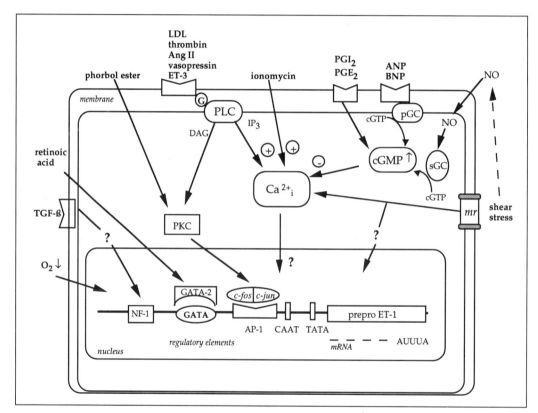

Fig. 2.3. Transcriptional regulation of the prepro ET-1 gene. See text for detailed discussion. Factors acting to control prepro ET-1 mRNA production include hypoxia ($O_2\downarrow$); tranforming growth factor beta (TGF-β); low density lipoprotein (LDL); angiotensin II (Ang II); endothelin-3 (ET-3); calcium ionophore (ionomycin); prostacyclin (PGI_2); prostaglandin E_2 (PGE_2); atrial natriuretic peptide (ANP); brain natriuretic peptide (BNP) and nitric oxide (NO). Shear stress, detected by mechanoreceptors (mr) influences ET synthesis both directly (solid line), and indirectly (dotted line) via NO (see text for details). The second messenger systems involved in regulation of ET production include G-proteins (G); phospholipase C (PLC); diacylglycerol (DAG); protein kinase C (PKC); inositol 1,4,5,triphosphate (IP_3); intracellular calcium concentration (Ca^{2+}_i); particulate guanylate cyclase (pGC); soluble guanylate cyclase (sGC); cyclic guanosine triphosphate (cGTP); cyclic guanosine monophosphate (cGMP). The circled symbols + and – represent stimulation and inhibition of Ca^{2+} release consecutively. Adapted from Hilkert et al, 1992.[10]

synthesis and dephosphorylation of c-*jun* in porcine endothelial cells supports this hypothesis.[24] **Insulin** increases ET-1 mRNA levels in bovine aortic endothelial cells via both PKC-dependent and -independent pathways.[33] It is postulated that insulin-responsive elements located in the prepro ET-1 gene mediate the PKC-independent pathway.[33]

The immunosuppressive agent **cyclosporine A** has been shown to stimulate ET synthesis in vitro,[34] an action which may partially explain its hypertensive effect in vivo.[35] Recent evidence suggests that this stimulation is dependent on release of transforming growth factor β,[36] implicating the NF-1 sequence in regulation of prepro ET-1 transcription by cyclosporine.

Exposure of rats to normobaric **hypoxia** for 24 hours results in increased ET-1 mRNA in the lungs and right atrium and increased plasma immunoreactive ET.[37] Hypoxic conditions also increase ET-1 release from cultured human endothelial cells,[38,39] and pulmonary arterial levels of ET-1 are increased in association with pulmonary hyperactivity during acute hypoxia in children with pulmonary hypertension.[40] Low oxygen tension induces expression of a number of genes and although the molecular mechanism is poorly understood it has been suggested that a heme protein may function as an oxygen tissue sensor, or that the redox state of certain transcription factors may function as second messengers.[41]

Low levels of **shear stress** (1.8 dyne/cm²) elevate ET-1 mRNA and immunoreactive ET-1 in cultured endothelial cells.[42-44] In human umbilical endothelial cells, this is proposed to be regulated by PKC,[44,45] although Morita et al[43] have shown that disruption of the actin cytoskeleton may also play a role. In contrast, higher levels of shear stress (>6 dyne/cm²) progressively reduce ET-1 mRNA transcription and release of immunoreactive ET-1.[44,46] Malek et al[46] found that neither the transcription factor AP-1, nor the GATA-2 binding site conferred shear responsiveness to the reporter gene, but instead implicated an alternative upstream *cis* element. Kuchan et al[44] linked inhibition of ET release by shear stress to enhanced release of **nitric oxide** (NO) from the endothelium, activation of smooth muscle soluble guanylate cyclase and subsequent accumulation of cyclic GMP. Inhibition of ET release by high shear stress was prevented by an inhibitor of NO synthase or by methylene blue, an inhibitor of guanylate cyclase and was mimicked, in stationary cells, by a stable analogue of cGMP, 8-bromo-cGMP.[44] Boulanger et al[47] were the first to describe inhibition of ET release by endothelium-derived NO, released during stimulation of porcine aortic endothelium with thrombin. Interestingly, atrial and brain **natriuretic peptides**, which increase cellular cGMP concentrations through activation of particulate guanylate cyclase, inhibit Ang II-stimulated ET-1 generation in rat mesangial cells.[30] This is despite the reported ability of atrial natriuretic peptide to increase the half-life of the prepro ET-1 transcript.[48] Cyclic GMP is implicated in the inhibition of ET-1 transcription, translation and secretion by the **prostanoids** prostaglandin E_2 (PGE$_2$) and prostacyclin (PGI$_2$).[49] Inhibition of ET-1 translation and secretion by PGE$_2$ and PGI$_2$ in bovine aortic endothelial and smooth muscle cells is prevented by an inhibitor of cGMP generation and by an inhibitor of cGMP-dependent protein kinase, but is unaffected by an inhibitor of cAMP-dependent protein kinase. Cyclic GMP might act to limit transcription by reducing the availability of intracellular calcium, because calcium chelation similarly reduces ET-1 liberation from endothelial cells.[21] Heparin inhibits thrombin-stimulated ET synthesis via NO,[50] but also reduces basal ET-1 release from endothelial cells by inhibition of a PKC-dependent pathway.[51] This mechanism seems to involve inhibition of prepro ET-1 mRNA expression following inhibition of c-*fos* proto-oncogene expression.[52] Retinoic acid also causes a down-regulation of ET-1 mRNA, probably via the GATA-2 factor.[14]

POST-TRANSCRIPTIONAL PROCESSING OF THE ET GENES

The 3' untranslated flanking region of prepro ET-1 mRNA contains 3 AUUUA motifs, similar to those of transiently expressed cytokines or growth factors and cellular proto-oncogenes.[20] These AU rich 'suicide motifs' may mediate selective destabilization of preproET-1 mRNA, accounting for its relatively short half-life of 15 minutes.[1,20] Yanagisawa et al[53] suggested that the prepro ET-1 mRNA instability might represent a 'protective' mechanism that counterbalances overexpression of ET-1.

The 2 kilobases of prepro ET-1 mRNA are encoded in 5 exons distributed over 7 kilobases of the genome.[3,20] The noncoding introns are removed by a precise splicing process allowing the coding sequence to be translated into protein. Alternative splicing of transcribed mRNA leads to the production of mRNA variants with subtle differences in structure from the normally transcribed mRNA. Deletion or addition of base-pairs can result in altered translational efficiency and in changes in the stability, transport, localization and activity of the derived peptides.[32] O'Reilly et al have reported the existence of a number of alternatively spliced mRNAs for ET-2[54] and ET-3[55] in human tissue. The splice variants differ in regions of mRNA encoding the carboxy terminal portions containing the putative sites for post-translational processing, which may result in no processing at all, or in direct processing to the mature ET. For this reason it has been proposed that alternative splicing may act as an on/off switch for ET production.

POST-TRANSLATIONAL PROCESSING OF PREPRO ET

Prepro forms of ET-1, 2 and 3 have hydrophobic amino acid residues at the amino terminal characteristic of a secretory signal sequence (Fig. 2.1).[1,7,8,20] The presence of this sequence suggests that the nascent ET precursors are transported across the internal cellular membrane, after which the signal peptide is cleaved, probably at Gly17-Ala18 to form pro ET precursors.[56]

Further processing of pro peptides to the mature peptides is considered to occur through a two stage process. In the case of ET-1(Fig. 2.1), the initial stage is proposed to consist of proteolytic cleavage at pairs of dibasic amino acids to release the 39 (porcine) or 38 (human) amino acid form, big ET-1. This is achieved by cleavage of pro ET-1 between the Lys51-Arg52 preceding and Arg92-Arg93 following the big ET-1 sequence.[4] This step is thought to be similar to the processing of other peptide hormones and may be dependent on one of the recently described proprotein convertases.[57,58] The final processing step, which releases ET-1 from the biosynthetic intermediate big ET-1, is unusual and requires the selective cleavage of the Trp21-Val22 bond in the carboxy terminal of big ET-1 (Fig. 2.2). The importance of this previously undescribed amino acid cleavage site implies the existence of a specific "endothelin-converting enzyme" (ECE).[1] Similar post-translational processing has been proposed for maturation of ET-2 and ET-3 precursors.

THE NATURE OF ECE

Several ECE-like enzyme activities representing different endopeptidase classes have been identified. Early evidence was provided suggesting that the putative ECE could be a **serine protease** with chymotrypsin-like activity.[1,59-61] Chymotrypsin cleaves big ET-1 to yield products which have in vitro and in vivo responses similar to ET-1.[59] A big ET converting enzyme with chymotrypsin-like activity was recently identified in the lipoprotein fraction of human serum, raising the possibility that big ET might be converted in the circulation.[62] Watanabe et al,[63] however, failed to find evidence for significant conversion of big ET-1 in circulating blood. In another recent study, furin, a novel serine protease was implicated in the processing of big ET-1 by bovine endothelial cells.[64] The products of big ET-1 hydrolysis by the **aspartic proteases** like pepsin,[65] cathepsin D [66] and others[67] also have ET-like activity. Processing of big ET by these enzymes is optimal at pH 4.0 and is inhibited by pepstatin A.[67,68] Membrane bound and cytosolic aspartic proteases with ECE-like activities have also been identified in several cells and tissues (reviewed in ref. 69). However, a major physiological role of these enzymes in biosynthesis of the ETs is doubtful for several reasons: the requirement of an acidic environment for optimal activity is not in accord with conversion of big ET at neutral pH; cleavage of the Asn18-Ile19 bond of big ET-1 in addition to Trp21-Val22 would result in simultaneous degradation of ET-1[66]; and pepstatin A does not affect ET-1 release from bovine endothelial cells, despite being present in sufficient quantity to inhibit all aspartic protease activity.[70] Deng et al[71] have partially purified an enzyme from the cytosol of porcine aortic endothelial cells with a pH optimum between

7 and 7.5 for conversion of big ET-1 to biologically active ET-1. The activity of this enzyme is inhibited by thiol protease inhibitors, but not by serine or metalloprotease inhibitors suggesting that it is a **soluble thiol protease.**

The majority of accumulated evidence suggests that the physiologically relevant ECE is likely to be a **metalloprotease.**[69] These enzymes have a narrow neutral pH optimum (6.8) and are inhibited by the chelating agents EDTA and 1,10-phenanthroline, and by low concentrations of divalent cations (Cu^{2+}, Zn^{2+}, Co^{2+}, Fe^{2+}).[68,72-74] Inactivation of ECE by EDTA is completely reversed by addition of Zn^{2+} suggesting the enzyme may be a zinc containing metalloprotease.[73] The predominant metalloprotease ECE is membrane bound and can be inhibited by the neutral (metallo) endopeptidase (NEP 24.11) inhibitor, phosphoramidon.[68,72-74] The activity of this ECE is not affected by other inhibitors of NEP, thiorphan or kelotorphan, or by inhibitors of the neutral metalloprotease, angiotensin-converting enzyme. Two other minor forms of metalloprotease enzyme with ECE activity have been described, both of which are soluble. One of them is sensitive[75] and the other insensitive[76] to inhibition by phosphoramidon.

Inhibition of ET-1 release from cultured endothelial cells[74] and marked reduction of the regional and systemic effects of big ET-1 in vivo by phosphoramidon, but not thiorphan (chapter 6),[77-79] demonstrates the physiological relevance of the membrane-bound metalloprotease. Similarly, phosphoramidon inhibits conversion of big ET-3 to ET-3 by endothelial cells[80] and the systemic effects of intravenously administered big ET-2 and big ET-3 in anesthetized rats,[81,82] raising the possibility that all of the big ETs are processed through a similar phosphoramidon-sensitive pathway. There is, however, some limited evidence supporting a physiological role of alternative processing enzymes. For example, the contractile effect of big ET-1 is inhibited by phosphoramidon in arteries but not in veins,[83] suggesting tissue differences in processing pathways.

PURIFICATION AND CLONING OF ET-1 SELECTIVE ECE

Our understanding of ET formation will undoubtedly be considerably improved by the recent purification, cloning and expression of membrane bound metalloprotease ECE. Membrane bound ECE was initially purified from rat endothelial cells[84] and lung[75] by exploiting the high affinity of this form of the enzyme for *Ricinus communis* agglutinin.[85] Shimada et al[86] and Xu et al[87] have subsequently cloned and expressed membrane bound ECE from rat endothelium and bovine adrenal cortex. Screening of cDNA libraries from other tissues including bovine adrenal cortex, bovine lung and cultured endothelial cells reveals several clones of the ECE cDNA, all of which appear to be derived from the same mRNA species.[87] The cDNA from rat endothelial cells encodes a polypeptide with 754 amino acids,[86] while that from bovine adrenal predicts a 758 amino acid sequence.[87] Comparison of the deduced amino acid sequences reveals a number of shared characteristics. Both sequences contain 10 potential glycosylation sites. A high level of glycosylation is consistent with the difference between the molecular mass predicted from the amino acid sequence and the actual mass of the purified enzymes (120-130kD). Exposure of purified ECE to the deglycosylating agent, endo-β-acetylglucosamidase F reduces the molecular weight to the anticipated 86kD.[86] Both sequences also contain a single putative transmembrane domain, predicted to be between residues 53-73 from the endothelial cDNA and between residues 57-78 from the bovine adrenal cortex cDNA. The large extracellular C-terminal of both enzymes contains a consensus sequence for a zinc binding motif that is common to the catalytic domains of many metalloproteases. Comparison of the deduced sequences with those in databases reveals a high degree of identity between ECE and both rat neutral endopeptidase (NEP) and human Kell blood group protein, a putative neutral endopeptidase. NEP and Kell also contain a single transmembrane domain and are highly glycosylated. Conservation of 10 out of 14 cysteine residues between these proteins indicates that their tertiary structures are likely to be similar. The greatest sequence similarity between the proteins is within the catalytic C-terminal domain, including the zinc binding motif. These similarities provide an explanation for the efficacy of the NEP inhibitor phosphoramidon in inhibiting ET conversion by ECE.

Substrate Specificity

Given the high degree of structural similarity between the NEP and ECE-1 metalloproteases it is surprising to find that they have quite different cleavage site selectivity. Unlike NEP which rapidly degrades many small peptides, including ET-1 itself, by cleavage at multiple internal sites,[88] ECE-1 cleaves only the Trp21-Val22 bond of the big ET's.[87]

Most studies where substrate specificity has been examined showed that the membrane bound metalloprotease cleaves big ET-3 less well than big ET-1.[68,72,73] The rat lung ECE, recently purified and characterized by Takahashi et al[89] cleaves all 3 big ETs, the ratio of conversion rates for big ET-1, -2, and -3 being approximately 4:1:2.[89] When expressed in COS cells the cloned bovine adrenal cortex ECE, termed ECE-1, also processes big ET-1 more efficiently than big ET-2 or big ET-3 (Fig. 2.4),[87] while the cloned rat endothelial cell enzyme cleaves big ET-1 but not big ET-2 or big ET-3.[86]

The sequences of big ET-1, -2 and -3 are highly conserved around the cleavage site at residues 15-26, except that in big ET-3 the Val22 at the cleavage site is replaced by Ile (Fig. 2.2). However, alterations in this region seem unlikely to explain differential conversion rates of the big ET-1, -2 and -3 by ECE. In an elegant set of experiments,

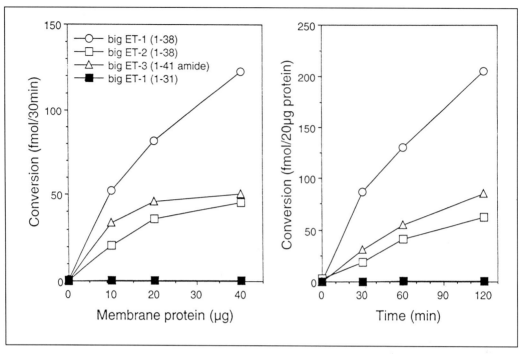

Fig. 2.4. Isopeptide selectivity of the endothelin-converting enzyme cloned from membrane fractions of bovine adrenal cortex and expressed in CHO cells. Synthetic ET precursor peptides, big ET-1, big ET-2, big ET-3 and the truncated big ET-1 (1-31) are used as substrates for the enzyme expressed in solubilized CHO/ECE membranes. Figure kindly provided by Professor M. Yanagisawa and reproduced with permission from Xu et al, Cell 1994; 78: 478-485 (copyright Cell Press, Massachusetts).

Fabbrini et al[90] expressed a set of prepro ET-1 mutants in *Xenopus* oocytes. Four mutants expressing alterations of the Trp21-Val22 site in pro ET-1 produced equivalent amounts of ET-1 as oocytes expressing wild type pro ET-1. This result implies that there is only a conformational requirement for cleavage at this site. It has been suggested that the tertiary structure of the substrate, particularly in the linear C-terminal region,[84,91] is a more important determinant of ECE activity.[59] The C-terminal residues 27-34 are different for big ET-2 and big ET-3 compared to big ET-1 (Fig. 2.2). Okada et al[91] proposed that these amino acid residues play a particularly important role in recognition of the substrate by the enzyme. This seems to be borne out by recent studies showing that deletion of the C-terminal residues 27-34 markedly reduces big ET-1 hydrolysis by purified ECE from endothelial cells[74] and that C-terminal truncated big ET-1(1-31) is not a substrate for the cloned ECE-1.[87] Interestingly, hydrolysis rates of linear big ET-1 (16-37) and big ET-2 (18-34) by purified endothelial ECE are more than twofold higher than the full length propeptides, suggesting that the amino terminal cyclic structures of big ET-1 and -2 might interfere with recognition by the enzyme.[84] The question of whether there are isoenzymes specific for each big ET still remains open. The majority of evidence to date supports processing of all three propeptides through a similar pathway, albeit with different apparent affinity.

CELLULAR LOCALIZATION OF ECE

Phosphoramidon-sensitive ECE is known to be almost entirely membrane bound,[73] but whether endogenous big ET-1 is cleaved by ECE bound to intracellular organelles or to plasma membrane is less clear. The fact that exogenous big ET-1 is converted to ET-1 both in vivo[79] and in cultured cells and isolated tissues in vitro[96] is supportive of an accessible plasma membrane binding site for ECE. Also, the presence of both big ET-1 and mature ET-1 in circulating blood and in the medium bathing cultured endothelial cells, suggests that the final conversion site might be at the plasma membrane. In the medium of cultured endothelial cells, over 50% of detectable ET is mature ET-1, with the remainder secreted as big ET-1.[92] However, conversion of exogenous big ET-1 by cultured endothelial cells is relatively inefficient, only 5-10% being converted to mature peptide.[69] It seems unlikely that such an inefficient mechanism is responsible for the final processing of big ET-1 physiologically.

Several lines of evidence support the existence of ECE activity within the cell. Antisera directed against both big ET-1 and ET-1 stain positively within the cytoplasm of endothelial cells[93] and mature ET-1 can be identified within a low density intracellular fraction of endothelial cells subjected to sucrose gradient centrifugation.[94] Furthermore, in two separate studies a low density intracellular fraction, proposed to contain constitutive secretory granules, was able to convert big ET-1 to the mature peptide.[93,94] One of these studies, using cultured endothelial cells,[94] suggested that conversion was by a phosphoramidon-insensitive pathway while the other, using rat lung,[93] found that secretion of big ET-1 was inhibited by phosphoramidon but not by inhibitors of serine, aspartic or thiol proteases. Both studies proposed that big ET-1 was converted during its transit through the intracellular constitutive secretory pathways, especially the Golgi apparatus. This location for the enzyme is consistent with the predicted type II integral membrane protein structure of the recently cloned phosphoramidon-sensitive ECE.[86,87] A role of secretory vesicles in the transport and processing of big ET could provide a mechanism for the polarized abluminal release of ET-1 from endothelial cells,[95] since directional secretion of peptides is entirely dependent on vesicle recognition of transport pathways.

In an elegant set of experiments Xu et al[87] showed that in COS cells co-transfected with the prepro ET-1 and ECE-1 genes 50-90% of the total ET peptide secreted is mature ET-1. In contrast, conversion of exogenous big ET-1 by ECE-1 expressing cells was only in the order of 5-10%. Thus conversion of endogenous big ET-1 is much more efficient than exogenous big ET-1. It has been estimated from kinetic studies of the purified phosphoramidon-sensitive ECE that, within physiological concentrations of the substrate, the rate of conversion is approximately proportional to the substrate concentration.[84,96] Xu et al[87] propose that endogenous big ET-1 is processed more efficiently because the concentration of big ET-1 achievable within secretory granules exceeds that achievable in the extracellular space. If this is the case then it is clear that the intracellular enzyme is of greater physiological significance than the plasma membrane bound enzyme. This might pose particular problems for the design of drugs that inhibit ECE. Intracellular conversion of big ET-1 by ECE-1 is inhibited much less potently by phosphoramidon and a novel ECE inhibitor, FR901533 (WS79089B)[97] than extracellular conversion, most likely because of the lesser accessibility of the intracellular site.[87]

ENDOTHELIN RELEASE: STORAGE OR DE NOVO SYNTHESIS?

Cultured endothelial cells take at least 30 minutes from stimulation with various vasoactive agents to increased release of ET into the surrounding milieu.[1] This time course predicts that ET release is regulated at the level of mRNA transcription/translation. In support of this, thrombin-stimulated ET release can be inhibited by the protein synthesis inhibitor cycloheximide.[47] These observations suggest that, for example in blood vessels, ET-1 is more likely to contribute to the long term control of vascular tone, in contrast to the rapid local control achievable by endothelium-derived relaxing factors. However, plasma immunoreactive ET-1 concentrations appear to increase in response to certain physiologic stimuli more rapidly than would be expected if ET-1 were always synthesized *de novo*. These stimuli include upright tilt in patients with vasovagal syncope[98] and the cold pressor test in healthy human subjects.[98] As discussed above, immunoreactive ET-1 has been detected in the cytoplasm of various cells. It is tempting to speculate that ET-1 formed in secretory granules[93,94] might provide a store that can be rapidly released in response to appropriate stimuli. Yoshizawa et al[99] showed that ET-like immunoreactivity in the posterior lobe of the rat pituitary gland was depleted after water deprivation, suggesting that stored ET-1 might play a role in neurosecretion.

TISSUE LOCATION OF ET AND ECE ACTIVITY

The ET-1 gene was first cloned from an endothelial cell cDNA library and, consequently, many of the above studies have been conducted using endothelial cells. However, application of in situ hybridization, radioimmunoassay and immunohistochemistry techniques have made it clear that expression of the ET family of genes, and ECE activity, are widespread. ET-1 is the major isoform synthesized by vascular endothelial cells. However, human endothelial cells may express the

gene for ET-2 in addition to that of ET-1 since big ET precursors of
both isopeptides are detectable in endothelial cells of human arteries
and veins.[100] The underlying smooth muscle cells also have the capacity to express the genes encoding the ET precursors[54,101] and to convert them to mature ETs.[74] Nonvascular smooth muscle cells of varying origin, including bladder[102] and airways,[103-105] release ET-1 and ET-2,
as do monocytes,[106] fibroblasts[102] and osteoblasts.[107] In the kidney, ET
mRNA and mature ET have been detected in association with mesangial
cells[30,108] and tubular epithelial cells.[109] ECE activity and ET-1 like
immunoreactivity can be detected in various regions of the rat brain,
particularly the hypothalamus, midbrain and medulla oblongata.[110]
Although this may be partially explained by endothelial production of
ET-1, there is evidence that cells of neural origin can secrete ETs.
ET-1 is released by cultured human astrocytes[111,112] and can be detected in association with parasympathetic ganglia in human airways.[105]
Firth and Ratcliffe[113] compared the tissue distribution of ET-1, ET-2
and ET-3 mRNAs in the rat. They found that ET-1 and ET-3 were
expressed in every tissue studied, with concentrations of ET-1 mRNA
being particularly high in the lung and large intestine and ET-3 mRNA
high in the small and large intestines, lung and kidney. In contrast,
ET-2 mRNA distribution was much more limited, with particularly
high levels being found in the small and large intestines and only low
levels in the stomach, skeletal muscle and heart. ET-1 and ET-3 were
also predominant in peripheral and CNS tissues of the pig.[114] Northern blot analysis of ECE-1 mRNA showed wide distribution of the
enzyme in bovine tissues, with particularly high expression in the lung,
adrenal gland, ovary and testis.[87] This may reflect the high level of
vascularization of these tissues since in situ hybridization using the
same probe showed that the most intense labeling was over the vascular endothelial cells of most tissues. Lower levels in the kidney and
brain are notable exceptions to this. Diffuse staining was also seen
over cardiac myocytes and hepatocytes, but surprisingly no staining
was seen over neuronal cells. The lack of hybridization particularly in
the neurones, known to be endothelin-producing cells, raises the possibility of other ECE isoenzymes.

ET-Like Peptides

Examination of the sequence of porcine prepro ET-1 from Cys 110
to Cys 124 reveals significant (53%) homology to the amino terminal
15 residues of ET.[1] Not only is this sequence flanked by pairs of dibasic
amino acids but the relative positions of the four cysteine residues are
conserved, raising the possibility that this 'ET-like peptide' might be
secreted and possess some biological activity. Similar 'ET-2 like' and
'ET-3 like' peptide sequences can be identified in the carboxy terminal portions of the prepro ET-2 and ET-3 genes.[7,8] An 'ET-1 like'
peptide is also encoded by the human ET-1 gene[3] and an 'ET-1 like'

sequence has been detected in human endothelial cells. Several studies have, however, failed to demonstrate any biological significance of the 'ET-1 like' peptide. It has no constrictor activity in porcine coronary artery[1] and does not inhibit the binding of [125I] ET-1.[115]

CLEARANCE AND DEGRADATION OF THE ETS

Intravenously administered [125I] ET-1 has a plasma half-life of less than 1 minute in anesthetized rats, most of the ET-1 being rapidly taken up by the lungs and kidneys.[116-118] In man, ET-1 can be cleared from the circulation by uptake in the lungs,[119] kidney and limbs,[120] although there are indications that pulmonary clearance may be less important in man than in other species.[121] Once taken up, ET seems to be concentrated in the cell membrane/internal organelle tissue fraction,[116] suggesting that clearance of ET-1 is by binding and then internalization.[122] It is currently unclear whether ET is removed from the circulation by binding to functional ET receptors or to specialized clearance receptors; an issue which is discussed further in chapter 4. Internalization of ET attached to receptors would allow degradation to be carried out within the cell. A possible candidate for an intracellular degrading enzyme is a soluble protease found in human platelets, vascular smooth muscle and endothelial cells.[123,124] This enzyme has a pH optimum of 5.5 and degrades ET rapidly by removal of the carboxyl terminal tryptophan, known to be important for binding of ET to its receptors.[125] A deamidase enzyme with similar characteristics was recently purified from rat kidney.[126,127]

ET may also be degraded without the necessity for internalization. Membrane bound neutral endopeptidases (EC 24.11), which are associated with venous and arterial endothelial cell plasma membranes can also degrade ET-1.[128] These enzymes are inhibited by phosphoramidon and SQ-28,603, both of which also inhibit the metalloprotease ECE,[128-130] and by thiorphan, which does not inhibit ECE.[86] The biological significance of this enzyme is not however clear, because inhibitors such as thiorphan do not potentiate the systemic effects of intravenously administered ET-1.[81] Activated polymorphonuclear lymphocytes are able to rapidly inactivate ET-1 through release of a protease, believed to be cathepsin G, which degrades ET by cleavage of His16-Leu17.[131,132] This process may have a role in acute inflammation where polymorphonuclear lymphocytes adhere to vascular endothelial cells.

FUTURE DIRECTIONS

The recent cloning of ECE was a major step forward in understanding the biosynthesis of the endothelins. Studies of ECE-1 distribution already suggested that isoenzymes of the enzyme might exist in neuronal tissue. Future work may identify other isoenzymes and allow more precise localization of specific ECE activities. Further analysis of the gene should also identify possible sites for transcriptional regulation of the ECE gene(s). Multiple sites for regulation of prepro ET-1 transcription have already been identified. Perhaps cellular synthesis of the ETs can be controlled at a second level by stimulation or inhibition of ECE transcription. This work might also help to unravel the mechanisms that allow rapid release of the ETs from some cells.

Progress has been made in the development of selective inhibitors of ECE activity.[97,133-135] Given the possible therapeutic applications of selective ECE inhibitors we should expect to see further developments in this area in the near future.

REFERENCES

1. Yanagisawa M, Kurihara H, Kimura S et al. A novel potent vasoconstrictor peptide produced by vascular endothelial cells. Nature 1988; 332: 411-415.
2. Itoh Y, Yanagisawa M, Ohkubo S et al. Cloning and sequencing of cDNA encoding the precursor of human endothelium-derived constrictor peptide, endothelin: identity of human and porcine endothelin. FEBS Lett 1988; 231: 440-444.
3. Bloch K, Freidrich SP, Lee M-E et al. Structural organisation and chromosomal assignment of the gene encoding endothelin. J Biol Chem 1989; 264: 10851-10857.
4. Yanagisawa M, Inoue A, Ishikawa T et al. Primary structure, synthesis and biological activity of rat endothelin, an endothelium-derived vasoconstrictor peptide. Proc Natl Acad Sci USA 1988; 85: 6964-6968.
5. Inoue A, Yanagisawa M, Kimura S et al. The human endothelin family: three structurally and pharmacologically distinct isopeptides predicted by three separate genes. Proc Natl Acad Sci USA 1989; 86: 2863-2867.
6. Hoehe MR, Ehrenreich H, Otterud B et al. The human endothelin-1 gene (EDN1) encoding a peptide with potent vasoactive properties maps distal to HLA on chromosome arm 6p in close linkage to D6S89. Cytogenet Cell Genet 1993; 62: 631-635.
7. Bloch KB, Hong CC, Eddy RL et al. cDNA cloning and chromosomal assignment of the endothelin 2 gene: vasoactive intestinal contractor peptide is rat endothelin 2. Genomics 1991; 10: 236-242.
8. Bloch KD, Eddy RL, Shows TB et al. cDNA cloning and chromosomal assignment of the gene encoding endothelin 3. J Biol Chem 1989; 264: 18156-18161.
9. Saida K, Mitsui Y, Ishida N et al. A novel peptide, vasoactive intestinal constrictor, of a new (endothelin) peptide family. J Biol Chem 1989; 264: 14964-14959.
10. Hilkert RJ, Lee M-U, Quertermous T. Genetic regulation of endothelin-1 in vascular endothelial cells. Trends Cardiovasc Med 1992; 2: 129-133.
11. Lee M-E, Bloch K, Clifford J et al. Functional analysis of the endothelin-1 promoter: evidence for an endothelial cell-specific cis-acting sequence. J Biol Chem 1990; 10: 10446-10450.
12. Wilson D, Dorfman D, Orkin S. A non-erythroid GATA-binding pro-

tein is required for function of the human preproendothelin-1 promoter in endothelial cells. Mol Cell Biol 1990; 10: 4854-4862.

13. Lee M-E, Temizer D, Clifford J et al. Cloning of the GATA-binding protein that regulates endothelin-1 gene expresion in endothelial cells. J Biol Chem 1991; 266: 16188-16192.

14. Dorfman D, Wilson D, Bruns G et al. Human transcription factor GATA-2: evidence for regulation of preproendothelin-1 gene expression in endothelial cells. J Biol Chem 1992; 267: 1279-1285.

15. Benatti L, Bonecchi L, Cozzi L et al. Two prepro endothelin mRNAs transcribed by alternative promoters. J Clin Invest 1993; 91: 1149-1156.

16. Lee M-E, Dhadly MS, Temizer DH et al. Regulation of endothelin-1 gene expression by Fos and Jun. J Biol Chem 1991; 266: 19304-19309.

17. Curran T, Franza B. Fos and Jun: the AP-1 connection. Cell 1988; 55: 395-397.

18. Mignotte V, Wall L, DeBoer E et al. Two tissue specific factors bind the erythroid promoter of the human porphobilinogen deaminase gene. Nucleic Acids Res 1989; 17: 37-54.

19. Takuwa N, Takuwa Y, Yanagisawa M et al. A novel vasoactive peptide endothelin stimulates mitogenesis through inositol lipid turnover. J Biol Chem 1989; 264: 7856-7861.

20. Inoue A, Yanagisawa M, Takuwa Y et al. The human preproendothelin-1 gene: complete nucleotide sequence and regulation of expression. J Biol Chem 1989; 264: 14954-14959.

21. Emori T, Hirata Y, Imai T et al. Cellular mechanism of thrombin on endothelin-1 biosynthesis and release in bovine endothelial cell. Biochem Pharmacol 1992; 44: 2409-2411.

22. Hattori Y, Kasai K, Banba N et al. Effect of a phorbol ester on immunoreactive endothelin-1 release from cultured porcine aortic endothelial cells. Endocrinol Jpn 1992; 39: 341-345.

23. Kohno M, Yokokawa K, Horio T et al. Release mechanism of endothelin-1 and big endothelin-1 after stimulation with thrombin in cultured porcine endothelial cells. J Vasc Res 1992; 29: 56-63.

24. Kitazumi K, Tasaka K. The role of c-jun protein in thrombin-stimulated expression of preproendothelin-1 mRNA in porcine aortic endothelial cells. Biochem Pharmacol 1993; 46: 455-464.

25. Perreault T, Stewart DJ, Cernacek P et al. Newborn piglet lungs release endothelin-1: effect of alpha-thrombin and hypoxia. Can J Physiol Pharmacol 1993; 71: 31-40.

26. Mikkola T, Ristimaki A, Viinikka L et al. Human serum, plasma and platelets stimulate prostacyclin and endothelin-1 synthesis in human vascular endothelial cells. Life Sci 1993; 53: 283-289.

27. Boulanger CM, Tanner FC, Bea ML et al. Oxidized low density lipoproteins induce mRNA expression and release of endothelin from human and porcine endothelium. Circ Res 1992; 70: 1191-1197.

28. Emori T, Hirata Y, Ohta K et al. Cellular mechanisms of endothelin-1 release by angiotensin and vasopressin. Hypertension 1991; 18: 165-170.

29. Imai T, Hirata Y, Emori T et al. Induction of endothelin-1 gene by angiotensin and vasopressin in endothelial cells. Hypertension 1992; 19: 753-757.

30. Kohno M, Horio T, Ikeda M et al. Angiotensin II stimulates endothelin-1 secretion in cultured rat mesangial cells. Kidney Int 1992; 42: 860-866.

31. Saijonmaa O, Nyman T, Fyhrquist F. Endothelin-1 stimulates its own synthesis in human endothelial cells. Biochem Biophys Res Commun 1992; 188: 286-291.

32. Benatti L, Fabbrini MS, Patrono C. Regulation of ET-1 biosynthesis. Ann N Y Acad Sci 1994; 714: 109-121.

33. Oliver FJ, de la Rubia G, Feener E et al. Stimulation of endothelin-1 gene expression by insulin in endothelial cells. J Biol Chem 1991; 266: 23251-23256.

34. Takeda Y, Itoh Y, Yoneda T et al. Cyclosporine A induces endothelin-1 release from cultured rat vascular smooth muscle cells. Eur J Pharmacol 1993; 233: 2-3.

35. Grieff M, Loertscher R, Al Shohaib S et al. Cyclosporine-induced elevation in circulating endothelin-1 in patients with solid-organ transplants. Transplantation 1993; 56: 880-884.

36. Khanna A, Li B, Stenzel KH et al. Regulation of new DNA synthesis in mammalian cells by cyclosporine: demonstration of a transforming growth factor beta-dependent mechanism of inhibition of cell growth. Transplantation 1994; 577-582.

37. Elton TS, Oparil S, Taylor GR et al. Normobaric hypoxia stimulates endothelin-1 gene expression in the rat. Am J Physiol 1992; 263: R1260-R1264.

38. Gertler JP, Ocasio VH, Dalman RL et al. Endothelin production by hypoxic human endothelium. J Vasc Surg 1993; 18: 178-184.

39. Kourembanas S, McQuillan LP, Leung GK et al. Nitric oxide regulates the expression of vasoconstrictors and growth factors by vascular endothelium under both normoxia and hypoxia. J Clin Invest 1993; 92: 99-104.

40. Allen SW, Chatfield BA, Koppenhafer SA et al. Circulating immunoreactive endothelin-1 in children with pulmonary hypertension: association with acute hypoxic pulmonary vasoreactivity. Am Rev Respir Dis 1993; 148: 519-522.

41. Helfman T, Falanga V. Gene expression in low oxygen tension. Am J Med Sci 1993; 306: 37-41.

42. Yoshizumi M, Kurihara H, Sugiyami T et al. Hemodynamic shear stress stimulates endothelin production by cultured cells. Biochem Biophys Res Commun 1989; 161: 859-864.

43. Morita T, Kurihara H, Maemura K et al. Disruption of cytoskeletal structures mediates shear stress-induced endothelin-1 gene expression in cultured porcine aortic endothelial cells. J Clin Invest 1993; 92: 1706-1712.

44. Kuchan MJ, Frangos JA. Shear stress regulates endothelin-1 release via protein kinase C and cGMP in cultured endothelial cells. Am J Physiol 1993; 264: H150-H156.

45. Wang DL, Tang CC, Wung BS et al. Cyclical strain increases endothelin-1 secretion and gene expression in human endothelial cells. Biochem Biophys Res Commun 1993; 195: 1050-1056.

46. Malek AM, Greene AL, Izumo S. Regulation of endothelin-1 gene by fluid shear stress is transcriptionally mediated and independent of protein kinase C and cAMP. Proc Natl Acad Sci USA 1993; 90: 5999-6003.

47. Boulanger C, Lüscher TF. Release of endothelin from the porcine aorta: inhibition by endothelium-derived nitric oxide. J Clin Invest 1990; 85: 587-590.

48. Hu RM, Levin ER, Pedram A et al. Insulin stimulates production and secretion of endothelin from bovine endothelial cells. J Biol Chem 1992; 267: 17384-17389.

49. Prins BA, Hu R-M, Nazario B et al. Prostaglandin E_2 and prostacyclin inhibit the production and secretion of endothelin from cultured endothelial cells. J Biol Chem 1994; 269: 11938-11944.

50. Yokokawa K, Tahara H, Kohno M et al. Heparin regulates endothelin production through endothelium-derived nitric oxide in human endothelial cells. J Clin Invest 1993; 92: 2080-2085.

51. Imai T, Hirata Y, Emori T et al. Heparin has an inhibitory effect on endothelin-1 synthesis and release by endothelial cells. Hypertension 1993; 21: 353-358.

52. Imai T, Hirata Y, Marumo F. Heparin inhibits endothelin-1 and proto-oncogene c-fos gene expression in cultured bovine endothelial cells. J Cardiovasc Pharmacol 1993; 22, Suppl 8: S49-S52.

53. Yanagisawa M, Inuoe A, Takuwa Y et al. The human preproendothelin-1 gene: Possible regulation by endothelial phosphoinosotide turnover signaling. J Cardiovasc Pharmacol 1989; 13: S13-S17.

54. O'Reilly G, Charnock-Jones DS, Morrison JJ et al. Alternatively spliced mRNAs for human endothelin-2 and their tissue distribution. Biochem Biophys Res Commun 1993; 193: 834-840.

55. O'Reilly G, Charnock-Jones DS, Davenport AP et al. Presence of messenger ribonucleic acid for endothelin-1, endothelin-2 and endothelin-3 in human endometrium and a change in the ratio of ET_A and ET_B receptor subtype across the menstrual cycle. J Clin Endocrinol Metab 1992; 75: 1545-1549.

56. Masaki T, Kimura S, Yanagisawa M et al. Molecular and cellular mechanism of endothelin regulation: implications for vascular function. Circulation 1991; 84: 1457-1468.

57. Steiner DF, Smeekens SP, Ohagi S et al. The new enzymology of precursor processing endoproteases. J Biol Chem 1992; 267: 23435-23438.

58. Seidah NG, Day R, Marcinikiewicz M et al. Mammalian paired basic amino acid convertases of prohormones and proproteins. Ann N Y Acad Sci 1993; 680: 135-146.

59. McMahon EG, Fok KF, Moore WM et al. In vitro and in vivo activity of chymotrypsin-activated big endothelin (porcine 1-40). Biochem Biophys Res Commun 1989; 161: 406-413.

60. Takaoka M, Takenobu Y, Miyata Y et al. Mode of cleavage of pig big endothelin-1 by chymotrypsin. Biochem J 1990; 270: 541-544.

61. Kaw S, Hecker M, Vane JR. The two-step conversion of big endothelin-1 to endothelin-1 and degradation of endothelin 1 by subcellular fractions from human polymorphonuclear leukocytes. Proc Natl Acad Sci USA 1992; 89: 6886-6890.

62. Ohwaki T, Sakai H, Hirata Y. Endothelin-converting enzyme activity in human serum lipoprotein fraction. FEBS Lett 1993; 320: 165-168.

63. Watanabe Y, Naruse M, Manzeri C. Is big endothelin converted to endothelin-1 in circulating blood? J Cardiovasc Pharmacol 1991; 17, Suppl 8: S503-S505.

64. Laporte S, Denault JB, D'Orleans-Juste P et al. Presence of furin mRNA in cultured bovine endothelial cells and possible involvement of furin in the processing of the endothelin precursor. J Cardiovasc Pharmacol 1993; 22, Suppl 8: S7-S10.

65. Takaoka M, Takenobu Y, Miyata Y et al. Pepsin, an aspartic protease, converts porcine big endothelin to 21-residue endothelin. Biochem Biophys Res Commun 1990; 166: 436-442.

66. Sawamura T, Kimura S, Shinmi O et al. Purification and characterization of putative endothelin-converting enzyme in bovine adrenal medulla: evidence for a cathepsin D like enzyme. Biochem Biophys Res Commun 1990; 168: 1230-1236.

67. Knap AK, Soriaano A, Savage P et al. Identification of a novel aspartyl endothelin-converting enzyme in porcine aortic endothelial cells. Biochem Mol Biol Int 1993; 29: 739-745.

68. Matsumura Y, Ikegawa R, Tsukuhara Y et al. Conversion of big-endothelin-1 to endothelin-1 by two types of metalloproteinases derived from porcine aortic endothelial cells. FEBS Lett 1990; 272: 166-170.

69. Opgenorth T, Wu Wong JR, Shiosaki K. Endothelin-converting enzymes. FASEB J 1992; 6: 2653-2659.

70. Shields PP, Gonzales TA, Charles SJ et al. Accumulation of pepstatin in cultured endothelial cells and its effect on endothelin processing. Biochem Biophys Res Commun 1991; 177: 1006-1012.

71. Deng Y, Savage P, Shetty SS et al. Identification and partial purification of a thiol endothelin-converting enzyme from porcine aortic endothelial cells. J Biochem (Tokyo) 1992; 111: 346-51.

72. Ohnaka K, Takayanagi R, Ohashi M et al. Identification and characterisation of endothelin converting activity in cultured bovine endothelial cells. Biochem Biophys Res Commun 1990; 168: 1128-1136.

73. Okada K, Miyazaki Y, Takada J et al. Conversion of big endothelin-1 by membrane bound metalloendopeptidase in cultured bovine endothelial cells. Biochem Biophys Res Commun 1990; 171: 1192-1198.

74. Ikegawa R, Matsumura Y, Tsukuhara Y et al. Phosphoramidon, a metalloproteinase inhibitor, suppresses the secretion of endothelin-1 from cultured endothelial cells by inhibiting a big-endothelin converting enzyme. Biochem Biophys Res Commun 1990; 171: 669-675.

75. Takada J, Okada K, Ikenaga T et al. Phosphoramidon-sensitive endothelin-converting enzyme in the cytosol of cultured bovine endothelial cells. Biochem Biophys Res Commun 1991; 176: 860-865.

76. Matsumura Y, Ikegawa R, Tsukuhara Y et al. N-ethylamide differentiates endothelin converting activity by two types of metalloproteinases derived from vascular endothelial cells. Biochem Biophys Res Commun 1991; 178: 531-538.

77. Matsumura Y, Hisaki K, Takaoka M et al. Phosphoramidon, a metalloproteinase inhibitor, suppresses the hypertensive effect of big endothelin-1. Eur J Pharmacol 1990; 185: 103-106.

78. Fukuroda T, Noguchi T, Tsuchida S et al. Inhibition of biological actions of big endothelin-1 by phosphoramidon. Biochem Biophys Res Commun 1990; 172: 390-395.

79. McMahon EG, Palomo MA, Moore WM et al. Phosphoramidon blocks the pressor activity of big endothelin-1(1-39) in vivo and conversion of big endothelin-1(1-39) to endothelin-1(1-21) in vitro. Proc Natl Acad Sci USA 1991; 88: 701-703.

80. Matsumura Y, Tsukahara Y, Kuninobu K et al. Phosphoramidon-sensitive endothelin-converting enzyme in vascular endothelial cells converts big endothelin-1 and big endothelin-3 to their mature form. FEBS Lett 1992; 305: 86-90.

81. Pollock DM, Divish BJ, Milicic I et al. In vivo characterization of a phosphoramidon-sensitive endothelin-converting enzyme in the rat. Eur J Pharmacol 1993; 231: 459-464.

82. Mattera GG, Eglezos A, Renzetti AR et al. Comparison of the cardiovascular and neural activity of endothelin-1, -2, -3 and respective proendothelins: effects of phosphoramidon and thiorphan. Br J Pharmacol 1993; 110: 331-337.

83. Auguet M, Delaflotte S, Chabrier PE et al. The vasoconstrictor action of big endothelin-1 is phosphoramidon-sensitive in rabbit saphenous artery, but not in saphenous vein. Eur J Pharmacol 1992; 224: 101-102.

84. Ohnaka K, Takayanagi R, Nishikawa M et al. Purification and characterization of a phsphoramidon sensitive endothelin-converting enzyme in porcine aortic endothelium. J Biol Chem 1993; 268: 26759-26766.

85. Ohnaka K, Nishikawa M, Takayanagi R et al. Partial purification of phosphoramidon-sensitive endothelin converting enzyme in porcine aortic endothelial cells: high affinity for *Ricinus communis* agglutinin. Biochem Biophys Res Commun 1992; 185: 611-616.

86. Shimada K, Takahashi M, Tanzawa K. Cloning and functional expression of endothelin-converting enzyme from rat endothelial cells. J Biol Chem 1994; 269: 18275-18278.

87. Xu D, Emoto N, Giaid A et al. ECE-1: a membrane bound metalloprotease that catalyzes the proteolyic activation of big endothelin-1. Cell 1994; 78: 473-485.

88. Vijayaraghaven J, Scili AG, Carretero OA et al. The hydrolysis of endothelins by neutral endopeptidase 24.11 (enkephalinase). J Biol Chem 1990; 265: 14150-14155.

89. Takahashi M, Matsushita Y, Iijima Y et al. Purification and characterization of endothelin-converting enzyme from rat lung. J Biol Chem 1993; 268: 21394-21398.

90. Fabbrini MS, Vitale A, Pedrazzini E et al. In vivo expression of mutant preproendothelins: hierarchy of processing events but no strict requirement of Trp-Val at the processing site. Proc Natl Acad Sci USA 1993; 90: 3923-3927.

91. Okada K, Takada J, Arai Y et al. Importance of the C-terminal region of big endothelin-1 for specific conversion by phosphoramidon-sensitive endothelin converting enzyme. Biochem Biophys Res Commun 1991; 180: 1019-1023.

92. Sawamura T, Kasuya Y, Matsushita Y et al. Phosphoramidon inhibits the intracellular conversion of big endothelin-1 to endothelin-1 in cultured endothelial cells. Biochem Biophys Res Commun 1991; 174: 779-784.

93. Gui G, Xu D, Emoto N et al. Intracellular localization of membrane-bound endothelin-converting enzyme from rat lung. J Cardiovasc Pharmacol 1993; 22: S53-S56.

94. Harrison VJ, Corder R, Anggard EE et al. Evidence for vesicles that transport endothelin-1 in bovine aortic endothelial cells. J Cardiovasc Pharmacol 1993; 2: S57-S60.

95. Wagner OF, Christ G, Wojta T et al. Polar secretion of endothelin-1 by cultured endothelial cells. J Biol Chem 1992; 267: 16066-16068.

96. Okada K, Arai Y, Hata M et al. Big endothelin-1 structure is important for specific processing by endothelin-converting enzyme of bovine endothelial cells. Eur J Biochem 1993; 218: 493-498.

97. Tsurumi Y, Ohhata N, Iwamoto T et al. WS79089A, B and C, new endothelin converting enzyme inhibitors isolated from *Streptosporangium roseum*. No. 79089. J Antibiotics 1994; 47: 619-631.

98. Fyhrquist F, Saijonmaa O, Metsarrine K et al. Raised plasma endothelin-1 concentration following cold pressor test. Biochem Biophys Res Commun 1990; 169: 217-221.

99. Yoshizawa T, Shinmi O, Giaid A et al. Endothelin: a novel peptide in the posterior pituitary system. Science 1990; 247: 462-464.

100. Howard P, Plumpton C, Davenport A. Anatomical localization and pharmacological activity of mature endothelins and their precursors in human vascular tissue. J Hypertension 1992; 10: 1379-1386.

101. Anfossi G, Cavalot F, Massucco P et al. Insulin influences immunoreactive endothelin release by human vascular smooth muscle cells. Metab Clin Exp 1993; 42: 1081-1083.

102. De Tejada IS, Mueller JD, De las Morenas A et al. Endothelin in the urinary bladder. I. Synthesis of endothelin-1 by epithelia, smooth muscle and fibroblasts suggests autocrine and paracrine cellular regulation. J Urol 1992; 148: 1290-1298.

103. Marciniak SJ, Plumpton C, Barker PJ et al. Localization of immunoreactive endothelin and proendothelin in the human lung. Pulmonary Pharmacology 1992; 5: 175-82.

104. Springall DR, Howarth PH, Counihan H et al. Endothelin immunoreac-

tivity of airway epithelium in asthmatic patients. Lancet 1991; 337: 697-701.

105. McKay KO, Black JL, Diment LM et al. Functional and autoradiographic studies of endothelin-1 and endothelin-2 in human bronchi, pulmonary arteries, and airway parasympathetic ganglia. J Cardiovasc Pharmacol 1991; 17: S206-9.

106. Ehrenreich H, Rieckmann P, Sinowatz F et al. Potent stimulation of monocytic endothelin-1 production by HIV-1 glycoprotein 120. J Immunol 1993; 150: 4601-4609.

107. Sasaki T, Hong MH. Endothelin-1 localization in bone cells and vascular endothelial cells in rat bone marrow. Anat Rec 1993; 237: 332-337.

108. Kohan DE. Production of endothelin-1 by rat mesangial cells: regulation by tumor necrosis factor. J Lab Clin Med 1992; 119: 477-84.

109. Ong ACM, Jowett TP, Scoble JE et al. Effect of cyclosporin A on endothelin synthesis by cultured human renal cortical epithelial cells. Nephrol Dial Transplant 1993; 8: 748-753.

110. Warner TD, Budzik GP, Mitchell JA et al. Detection by bioassay and specific enzyme-linked immunosorbent assay of phosphoramidon-inhibitable endothelin-converting activity in brain and endothelium. J Cardiovasc Pharmacol 1992; 2446:

111. Jiang MH, Hoog A, Ma KC et al. Endothelin-1-like immunoreactivity is expressed in human reactive astrocytes. Neuroreport 1993; 4: 935-937.

112. Ehrenreich H, Costa T, Clouse KA et al. Thrombin is a regulator of astrocytic endothelin-1. Brain Res 1993; 600: 201-207.

113. Firth JD, Ratcliffe PJ. Organ distribution of the three rat endothelin messenger RNAs and the effects of ischemia on renal gene expression. J Clin Invest 1992; 90: 1023-1031.

114. Hemsen A, Lundberg JM. Presence of endothelin-1 and endothelin-3 in peripheral tissues and central nervous system of the pig. Regul Pept 1991; 36: 71-83.

115. Cade C, Lumma WC, Mohan R et al. Lack of biological activity of prepro endothelin(110-130) in several endothelin assays. Life Sci 1990; 47: 2097-2103.

116. Anggård E, Galton S, Rae G et al. The fate of radioiodinated endothelin-1 and endothelin-3 in the rat. J Cardiovasc Pharmacol 1989; 13, Suppl 5: S46-S49.

117. Shiba R, Yanagisawa M, Miyauchi T et al. Elimination of intravenously injected endothelin-1 from the circulation of the rat. J Cardiovasc Pharmacol 1989; 13, Suppl 5: S98-S101.

118. De Nucci G, Thomas R, D'Orleans-Juste P et al. Pressor effects of circulating endothelin are limited by its removal in the pulmonary circulation and by the release of prostacyclin and endothelium-derived relaxing factor. Proc Natl Acad Sci USA 1988; 85: 9797-9800.

119. Stewart DJ, Levy RD, Cernacek P et al. Increased plasma endothelin-1 in pulmonary hypertension: marker or mediator of hypertension? Ann Intern Med 1991; 114: 464-469.

120. Gasic S, Wagner OF, Vierhapper H et al. Regional hemodynamic effects

and clearance of endothelin-1 in humans: renal and peripheral tissues may contribute to the overall disposal of the peptide. J Cardiovasc Pharmacol 1992; 19: 176-80.

121. Ray SG, McMurray JJ, Morton JJ et al. Circulating endothelin is not extracted by the pulmonary circulation in man. Chest 1992; 102: 1143-1144.

122. Gandhi CR, Harvey SAK, Olson MS. Hepatic effects of endothelin: metabolism of [^{125}I] endothelin-1 by liver-derived cells. Arch Biochem Biophys 1993; 305: 38-46.

123. Jackman HL, Morris PW, Deddish PA et al. Inactivation of endothelin I by deamidase (lysosomal protective protein). J Biol Chem 1992; 267: 2872-2875.

124. Jackman HL, Morris PW, Rabito SF et al. Inactivation of endothelin-1 by an enzyme of the vascular endothelial cells. Hypertension 1993; 21: II 925-II 928.

125. Kimura S, Kasuya Y, Sawamura T et al. Structure-activity relationships of endothelin: importance of the C-terminal moiety. Biochem Biophys Res Commun 1988; 156: 1182-1186.

126. Deng AY, Martin LL, Balwierczak JL et al. Purification and characterization of an endothelin degradation enzyme from rat kidney. J. Biochem. 1994; 115: 120-125.

127. Janas J, Sitkiewicz D, Warnawin K et al. Characterization of a novel, high-molecular weight, acidic, endothelin-1 inactivating metallo-endopeptidase from the rat kidney. J Hypertension 1994; 22: 1155-1162.

128. Llorens-Cortes C, Huang H, Vicart P et al. Identification and characterisation of neutral endopeptidase in endothelial cells from venous or arterial origins. J Biol Chem 1992; 267: 14012-14018.

129. Abassi ZA, Tate JE, Golomb E et al. Role of neutral endopeptidase in the metabolism of endothelin. Hypertension 1992; 20: 89-95.

130. Abassi ZA, Golomb E, Bridenbaugh R et al. Metabolism of endothelin-1 and big endothelin-1 by recombinant neutral endopeptidase EC.3.4.24.11. Br J Pharmacol 1993; 109: 1024-1028.

131. Fagny C, Michel A, Nortier J et al. Enzymatic degradation of endothelin-1 by activated human polymorphonuclear neutrophils. Regul Pept 1992; 42: 1-2.

132. Patrignani P, Del-Maschio A, Bazzoni G et al. Inactivation of endothelin by polymorphonuclear leukocyte-derived lytic enzymes. Blood 1992; 78: 2715-2720.

133. Matsuura A, Okumura H, Asakura R et al. Pharmacological profiles of aspergillomarasmines as endothelin converting enzyme inhibitors. Jpn J Pharmacol 1993; 187-193.

134. Bertenshaw SR, Talley JJ, Rogers RS et al. Thiol and hydroxamic acid containing inhibitors of endothelin converting enzyme. Bioorg Med Chem Lett 1993; 1953-1958.

135. Morita A, Nomizu M, Okitsu M et al. D-Val^{22} containing human big endothelin-1 analog, (D-Val^{22})Big ET-1(16-38), inhibits the endothelin converting enzyme. FEBS Lett 1994; 353: 84-88.

======= CHAPTER 3 =======

MOLECULAR CHARACTERIZATION OF ENDOTHELIN RECEPTORS

Gillian A. Gray

Specific binding sites for [^{125}I] endothelin-1 (ET-1) have been identified in many tissues.[1,2] These binding sites can be classified according to their relative affinities for the ET isopeptides ET-1, ET-2 and ET-3. The ET$_A$ receptor is characterized by its very high (subnanomolar) affinity for ET-1 and ET-2 and its 70-100 fold lower affinity for ET-3 (Fig. 3.1). The ET$_B$ receptor in contrast, has high and equal affinity for all three isopeptides. These binding characteristics are reflected in the agonist potency of the isopeptides in functional studies (see chapter 4). Observations that ET-3 can be more a potent agonist than either ET-1 or ET-2 in pharmacological studies[3-5] have lead to speculation about the existence of a third, so called ET$_C$, receptor subtype. Complementary DNAs for ET$_A$, ET$_B$, ET$_C$ receptors and also an ET$_{AX}$ receptor subtype, have now been isolated and characterized, although only ET$_A$ and ET$_B$ subtypes have thus far been identified in the mammalian genome. All of the cDNAs encode proteins with 7 transmembrane spanning domains and other features common to the G-protein-linked superfamily of receptors (Fig. 3.2 and Fig. 3.3). The aim of this chapter is to consider the structures, regulation and distribution of the receptor genes, and the features which confer isopeptide selectivity on the receptors.

THE ET RECEPTOR GENES

In 1990, two years after the initial description of ET,[6] Arai et al[7] and Sakurai et al[8] succeeded in cloning the ET$_A$ and ET$_B$ receptor

Molecular Biology and Pharmacology of the Endothelins edited by Gillian A. Gray and David J. Webb. © 1995 R.G. Landes Company.

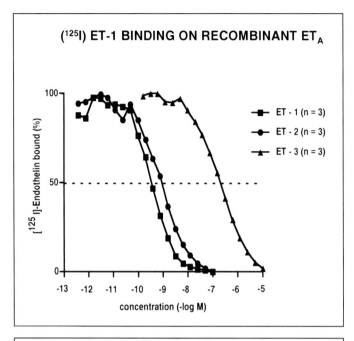

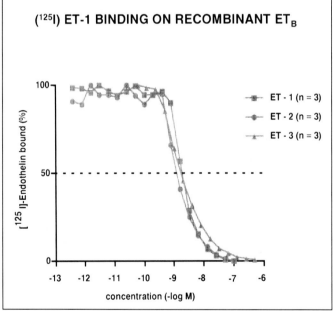

Fig. 3.1. Inhibition of [^{125}I]ET-1 binding to CHO cells expressing recombinant ET$_A$ receptor (upper panel) or recombinant ET$_B$ receptor (lower panel), cloned from human placenta.[12,37] Binding of [^{125}I]ET-1 to the ET$_A$ receptor is inhibited with greater potency by ET-1 and ET-2 than by ET-3, whereas binding to the ET$_B$ receptor subtype is inhibited with equal potency by all three ET isopeptides. Figures kindly provided by Dr. Volker Breu, Hoffman-La Roche Ltd, Basel, Switzerland.

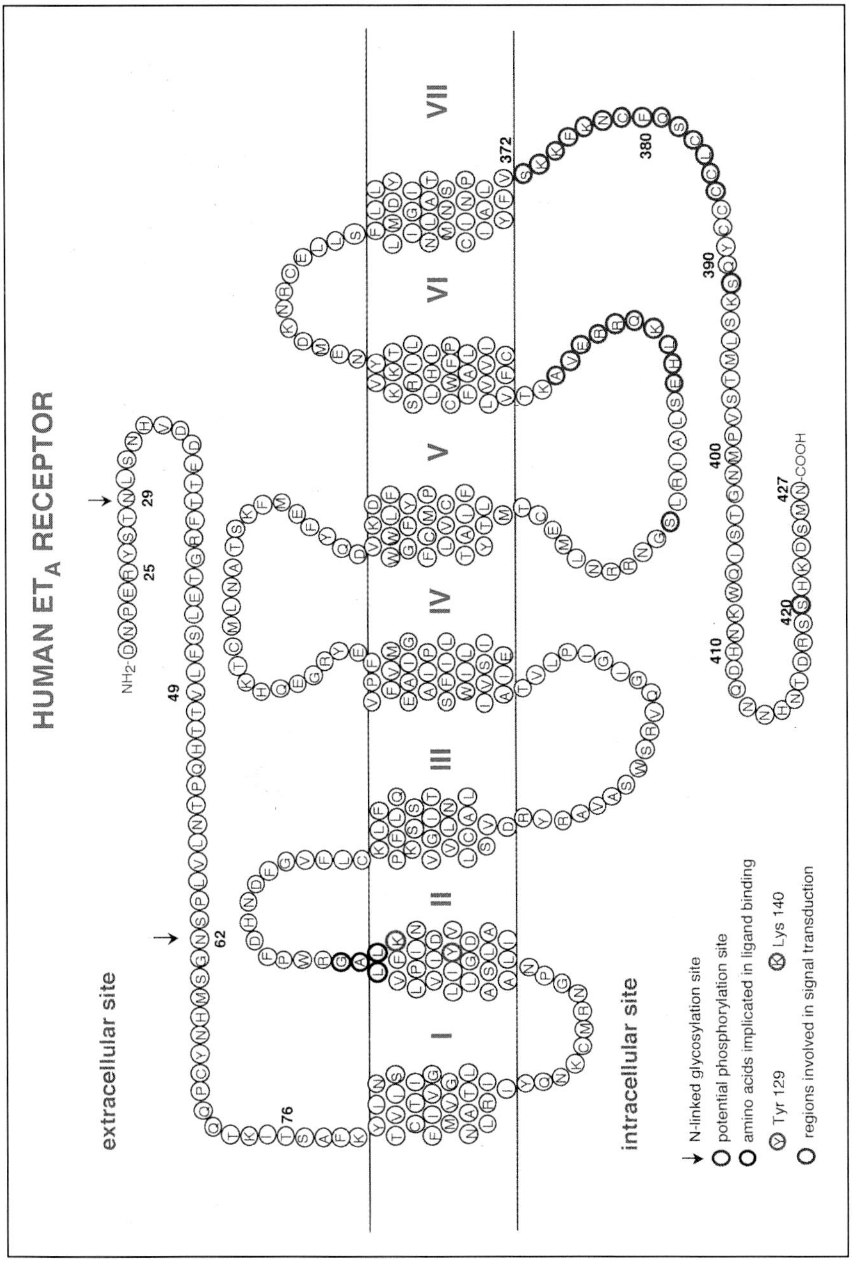

Fig. 3.2. The predicted structure of the mature form of the human ET_A receptor (after removal of the first 20 amino acids by a signal peptidase) including potential sites for post-translational modification (see text for details). The transmembrane domains are numbered I-VII. Figure modified from Adachi et al, 1993.[42]

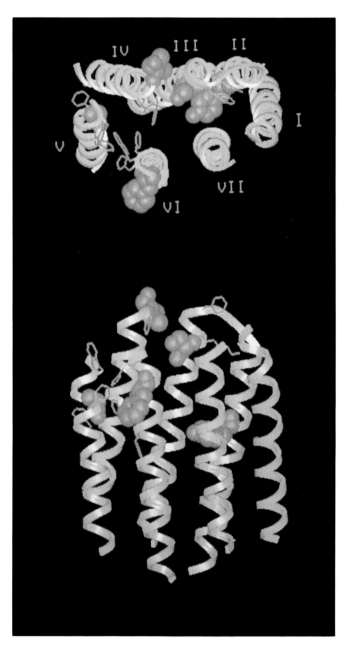

Fig. 3.3. Molecular model of the ET_A receptor predicted from amino acid sequence-derived hydrophobicity, based on the known coordinates of bacteriorhodopsin. upper: axial view of the extracellular face of the receptor model showing residues (green or purple) comprising the presumed ligand binding site. The residues in green are those targeted for site-directed mutation by Krystek et al.[48] Transmembrane helices are numbered I to VII. lower: cross-sectional view of the ET_A receptor model with the extracellular surface orientated toward the top. Figure kindly provided by Dr. S. Krystek and reproduced with permission from J Biol Chem 1994; 269: 12383-12386.

genes. The ET-1 selective ET_A receptor was cloned from a bovine lung cDNA library[7] and the nonisopeptide selective ET_B receptor from a rat lung cDNA library.[8] Shortly afterwards, Lin et al[9] reported cloning of an ET_A receptor from a rat vascular smooth muscle cell line. Subsequent screening of a human placental cDNA library by cross-hybridization with fragments of bovine ET_A cDNA or rat ET_B cDNA allowed the isolation of human ET_A and ET_B receptor cDNA clones.[10-12] Several other groups have since reported cloning and expression of human ET receptors from cDNA libraries of other tissues, including heart (ET_A[13,14] and ET_B[14]), lung (ET_A[15] and ET_B[13,15]), liver (ET_B[16]) and jejunum (ET_B).[17]

Analysis of genomic DNA shows that the ET_A and ET_B receptor genes have similar structural organization, suggesting that they originated from the same ancestral gene. The human ET_A receptor gene consists of 8 exons and 7 introns spanning more than 40 kilobases and is assigned to human chromosome 4.[18,19] The human ET_B receptor gene spans 24 kilobases consisting of 7 exons and 6 introns and is assigned to chromosome 13.[20] In both cases the coding region for the receptors is approximately 5 kilobases in length. The cDNAs for the human ET_A and ET_B receptors predict 427 and 442 amino acids respectively and the overall identity between the two proteins is reported to be between 55% and 64%, depending on the tissue studied.[20,13,21,12]

Low stringency Southern blot analysis of human genomic DNA with cDNA probes for the human ET_A and ET_B receptors identified only two hybridizing genomic DNA fragments.[17] This suggests that further ET receptor genes, if they exist in the human genome, should have quite low sequence similarities to the two known ET receptor genes. In the past year, two groups have identified alternative ET receptor clones by screening amphibian cDNA libraries. Karne et al[22] cloned an ET receptor from *Xenopus* dermal melanophores that shows relative selectivity for ET-3, consistent with the putative ET_C receptor subtype. The cDNA of this receptor encodes 424 amino acids and the amino acid sequence is 47 and 52% identical to bovine ET_A[7] and rat ET_B[8] receptors, respectively (Fig. 3.4).[22] The receptor cloned by Kumar et al[23] from *Xenopus* heart seemed initially to resemble the mammalian ET_A receptor in its relative selectivity for ET-1. However, binding of [^{125}I]ET-1 to this receptor, termed ET_{AX}, is not inhibited by the ET_A receptor selective ligand, BQ-123. The deduced amino acid sequence of the ET_{AX} receptor is 74, 72 and 74% identical to the human[13], bovine[7] and rat[9] ET_A receptor; 60 and 61% identical to the human[13] and porcine[24] ET_B receptor; and 51% identical to the *Xenopus* ET_C receptor.[22] Genes for these novel receptor subtypes have not yet been detected in the mammalian genome.

REGULATORY REGIONS OF THE ET_A AND ET_B RECEPTOR GENES

As with the ET genes (chapter 2), the nontranscribed 5' flanking regions of the ET receptor genes have been investigated in order to

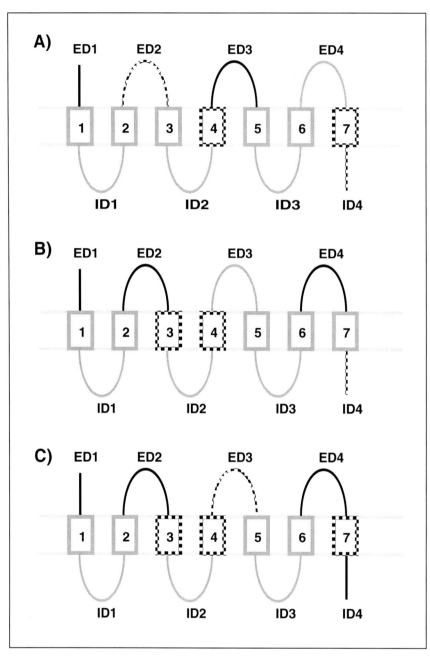

Fig. 3.4. Comparison of the amino acid sequences between the different domains of (A) ET_A^7 and ET_B^8 receptors (B) ET_A and ET_C^{22} receptors and (C) ET_B and ET_C receptors. The identity between the receptors is divided into 3 classes: ≥ 60% (grey lines), 30-60% (stipled lines), and ≤ 60% (black lines). ID, intracellular domain; ED, extracellular domain. The numbers in the boxes designate transmembrane domains. Figure kindly provided by Dr.M Lerner and reproduced with permission from J Biol Chem 1993; 268: 19126-19133.

characterize the regions regulating gene transcription. The ET_A and ET_B receptor genes contain a number of common factors including a consensus sequence for binding of the RNA polymerase II transcription factor, Sp-1[25] and another for a GATA motif which is recognized by DNA binding proteins like nuclear factor-1.[18,20] The GATA motif is also known to be important for the efficient expression of the human prepro-ET-1 gene.[26,27] Like the prepro ET-1 gene, the human ET_B receptor gene contains an acute phase reactant regulatory element, which confers sensitivity to acute physiological stress in vivo.[28] Both ET receptor genes contain several hexanucleotide sequences for 'E boxes', which interact with the basic helix-loop-helix transcription factor family. The ET_A receptor gene also contains a cluster of GAGA repeats in intron 1 which are suggested to play a role in stimulation of receptor expression.[19] Arai et al[20] reported that the human ET_B receptor gene does not contain either a TATA box which is thought to fix the site of transcription initiation,[29] or a CAAT box, which is usually located close to the transcription initiation site. Similarly, Hosada et al[10,18] found these elements to be missing from the human ET_A receptor gene. Both groups suggested that the lack of these regulatory regions might explain the presence of multiple initiation sites for ET_A and ET_B receptor transcription. The bovine ET_B receptor gene is also reported to have multiple initiation sites with no apparent TATA box sequence.[30] However, Yang et al[19] have recently reported the presence of a 'TATA-like' sequence in the 5' flanking region of the human ET_A receptor gene, close to the transcription initiation site.

The 3' untranslated flanking regions of both human ET_A[18] and ET_B[11,20] receptors contain potential polyadenylation signals that may mediate selective destabilization of the receptor mRNA.

POST-TRANSLATIONAL PROCESSING OF THE ET RECEPTORS

The first 20 N-terminal amino acids encoded by the ET_A, ET_B and ET_C receptor genes constitute a signal sequence that is cleaved by a signal peptidase on secretion of the receptors from the nucleus.[10,16,22]

The genes of all the cloned receptors encode several additional sites for post-translational modification that influence the tertiary structure of the receptor, its anchoring to the cell membrane and its

linkage to intracellular messenger systems (Fig. 3.3). These include consensus sites for N-glycosylation, several potential sites for palmitoylation to anchor the receptor to the cell membrane, and serine residues that may be substrates for regulatory phosphorylation by serine threonine kinases.[10,11,13,16]

STRUCTURAL REQUIREMENTS FOR LIGAND BINDING AND SELECTIVITY

All of the cloned ET receptor genes predict a heptahelical structure common to members of the G-protein-coupled receptor superfamily and similar to many neuropeptide receptors (Fig. 3.2 and Fig 3.3).[31] The regions of greatest sequence conservation between the ET receptors and other G-protein coupled receptors are concentrated within the hydrophobic transmembrane segments, whereas the amino and carboxyl-terminal regions and the loops between the hydrophobic segments IV, V and VI differ widely in length and amino acid composition.

Among the ET receptors, the 7 transmembrane domains and cytoplasmic loops of the receptors are highly conserved but the N-terminal and other extracellular domains exhibit differences in both length and amino acid sequences (Fig. 3.4). The homology between human ET_A and ET_B receptors in the N-terminal region is only 4%.[11,13] The ligand binding domains of several G-protein-coupled receptors are embedded deep within the plasma membranes.[32] However, for peptide G-protein-coupled receptors the extracellular N-terminal regions of the receptors are also known to be important in binding.[33] The differences exhibited between ET receptors in this region suggest that the N-terminal amino acids may play a role in either ligand binding or ligand selectivity.

The roles of the N-terminus, and other regions of the ET receptors, have been investigated by application of molecular biology techniques which allow substitution of single amino acids (site-directed mutagenesis), exchange of amino acid segments between receptors to form recombinant chimeric receptors, and truncation of the receptor proteins. Expression of these mutated receptors in heterologous cell lines like COS-7 cells and chinese hamster ovary (CHO) cells has allowed them to be characterized by competitive radioligand binding assays and agonist-induced intracellular calcium transient assays. Although this type of experiment can provide useful information, the results must be interpreted with some caution. It cannot be ruled out that mutations of the amino acids in one region of the receptor protein induce conformational changes that influence binding in a different region.

N-Terminus

Cheng et al[34] have identified alternative transcripts of the ET_B receptor in the rat brain which have four amino acid substitutions in the initial segment of the N-terminal sequence, but no alteration in binding affinity or specificity. Binding is also reported to be unaffected by proteolysis of the bovine lung ET_B receptor at a metalloprotease cleavage site between Asp79 and Gly80 that removes 79 of the 101 N-terminal amino acids.[35] Similarly, truncation of the initial 49 of the 80 N-terminal amino acid residues in the human ET_A receptor sequence has no influence of binding of [^{125}I] ET-1 relative to the wild type receptor.[36] In contrast, deletion of amino acids 25-76 of the human ET_A receptor (Fig. 3.2), which lie in close proximity to the first transmembrane region cause complete loss of [^{125}I] ET-1 binding.[36] Together these results suggest that if the N-terminal region of the ET receptors does influence ligand binding, the amino acids located nearest to the first transmembrane domain are those with which the ligand might interact. Site-directed mutagenesis suggests that Asp75 and Pro93, two amino acids located in this region of the human ET$_B$ receptor, are involved in the formation of a super-stable complex between the ET-1 and the receptor.[39] Once formed, this complex permits ET-1 to remain bound to the ET$_B$ receptor in the presence of the ionic detergent sodium dodecyl sulphate (SDS) and under acidic conditions.[37] Neither Asp or Pro are found at the appropriate position in the ET$_A$ receptor[38] and, although ET$_A$ receptors bind ET-1 stably and with similar affinity as the ET$_B$ receptor, ET-1 can easily be detached from them under acidic conditions.[39]

TRANSMEMBRANE DOMAINS

The second and third transmembrane domains of many G-protein-coupled receptors contain highly conserved aspartic acid residues that are considered to be important for ligand binding.[32,39] In the third transmembrane domain of both ET receptors, one of these is replaced by a lysine residue. Substitution of this Lys181 with Asp in the ET$_B$ receptor reduces the affinity of the receptor for the ET isopeptides without altering its G-protein coupling ability.[40] Interestingly, the affinity of the mutant receptor is reduced more for ET-3 and the ET$_B$ receptor selective ligand sarafotoxin S6c (SRTX S6c) than for ET-1 and ET-2. This Lys might therefore contribute not only to ligand binding but also to determination of ligand selectivity of the ET$_B$ receptor.

Adachi et al[41,42] have identified a region at the boundary of the second transmembrane domain of the ET$_A$ receptor that seems essential for ligand binding. Inhibition of [^{125}I] ET-1 binding to the ET$_A$ receptor by the ET$_A$ selective ligand, BQ-123,[43] is prevented when the first extracellular loop of the human ET$_A$ receptor is replaced by the corresponding loop of the ET$_B$ receptor. In contrast, introduction of the first extracellular loop of the ET$_A$ receptor into the ET$_B$ receptor endows the ET$_B$ receptor with sensitivity to BQ-123. Interestingly, the BQ-123 insensitive ET$_{AX}$ receptor shares only 58% homology with the ET$_A$ receptor in this first extracellular domain.[23] Adachi et al[45] have gone on to identify a series of five amino acids (140-144, Fig. 3.2) located on the border of the second transmembrane and first extracellular loop domains that seem essential for ET-1 binding to the ET$_A$ receptor. Substitution of the corresponding domain of the β_2-adrenergic receptor results in loss of BQ-123 binding to the ET receptor. Most recently, the same authors have utilized site-directed mutagenesis to identify the Lys140 residue as being particularly important for ligand binding to the ET$_A$ receptor.[45] Lys140 is conserved among ET receptors but not among other G-protein coupled receptors. The authors speculate that binding of ET-1 to Lys140, predicted to be located in the inner region of the receptor, changes the conformation of the ET_A receptor so that it may couple with a G-protein. Replacement of Gly^{144} also slightly reduced ET-1 binding, perhaps because it lies conformationally close to Lys140.

The amino acid sequences contained in transmembrane domains IV to VI and the intervening loops have recently been implicated as the principle determinants of ET receptor ligand selectivity.[42,45-47] ET_B-like binding characteristics are conferred on the ET_A receptor by substitution of transmembrane regions IV, V and VI and their intervening loops with the corresponding regions of the ET_B receptor (Fig. 3.5).[42,46] Interestingly, binding of labeled ET-1 and ET_B selective ligands to this 'ET$_B$-like chimera' is inhibited both by ET_A selective (BQ-123) and ET_B selective ligands (BQ-3020, IRL 1620 and ET-3, Fig. 3.5).[46,47] On the basis of these results, Sakamoto et al[46] suggest that the ET receptors can be divided into two distinct parts, one of which determines isopeptide selectivity (transmembrane domains IV, V and VI) and the other which is involved in ligand receptor binding (transmembrane domains I, II, III and VII).

Recent studies have used computer assisted molecular modeling, based on the known coordinates of bacteriorhodopsin, to identify nonconserved amino acids in the ET_A and ET_B receptors that may be important in determination of ligand selectivity (Fig. 3.3). Two studies identified Tyr129 in the second transmembrane domain as being of potential importance for binding of selective ligands to the ET_A receptor.[48,49] Replacement of this amino acid by alanine results in reduced affinity for the ET_A selective ligands BQ-123 and BMS-182874 and parallel increased affinity for the ET_B selective ligands ET-3 and SRTX S6c.[48] Interestingly, replacement of the analogous sequence in the human ET_B receptor by alanine has no influence on ligand binding.[49] Thus, although transmembrane domain II seems important in ligand selectivity for the human ET_A receptor, it does not appear to play a significant role in the lack of ligand selectivity shown by the human ET_B receptor.

CYTOPLASMIC LOOPS

The predicted third cytoplasmic domain of the ET receptors is very short (approximately 30-50 residues), a feature common to G-protein coupled receptors which have peptide ligands.[13] In β_2-adrenergic receptors the corresponding region interacts with G-proteins.[50] This domain of the human ET_A receptor has therefore been investigated to examine its ability to couple with G-proteins. Coupling of the ET receptor with G proteins in CHO cells can be evidenced by the ability of ET to evoke an increase in intracellular Ca^{2+} concentration.[42,51] In CHO cells expressing the human ET_A receptor, deletion of 16 of the amino acids in the third cytoplasmic domain reduces both ligand binding activity and coupling to calcium release.[51] The authors suggest that the deletions cause conformational changes in the receptor resulting in reduced binding and prevention of G-protein coupling. Using a slightly different approach, Adachi et al[42] substituted 10 amino acids at the C-terminal end of the third cytoplasmic domain in the ET_A receptor with the corresponding region of the β_2-adrenergic receptor (Fig. 3.2). This mutation has no effect on ligand binding activity, but almost abolishes the ability of ET-1 to increase intracellular

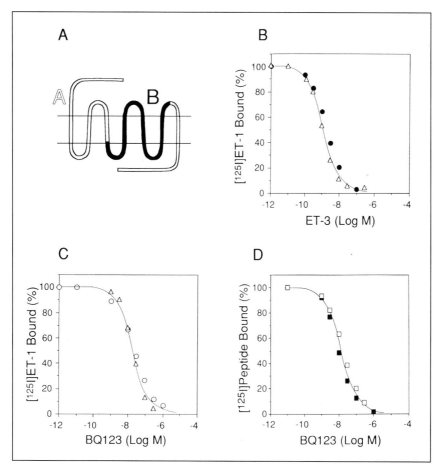

Fig. 3.5. Characterization of human ET_B-like chimeric receptor, expressed in Ltk⁻ cells. The white region derives from the ET_A receptor (A) and the black region from the ET_B receptor (B). (A) A schematic diagram of the ET_B-like chimera, (B) Inhibition of [¹²⁵I]ET-1 binding to a wild-type ET_B receptor (●) and to the ET_B-like chimera (Δ) by ET-3, (C) Inhibition of [¹²⁵I]ET-1 binding to the wild type ET_A receptor (○) and the ET_B-like chimera(Δ) by BQ123, (D) Inhibition of [¹²⁵I]ET-1 binding to the ET_B-like chimera by ET_B selective ligands [¹²⁵I]-BQ 3020 (□) and [¹²⁵I]-IRL 1620 (■) (see text for details). Figure kindly provided by Dr. A. Sakamoto and reproduced with permission from J Biol Chem 1993; 268:8547-8553.

calcium. In contrast, deletion of the N-terminus of the third intracellular domain does not change either ligand binding or the ability of ET-1 to evoke a Ca^{2+} signal.[42] Interaction of the receptor with the G-protein is thus most likely to be via the C-terminal part of this domain (Fig. 3.2).

C-TERMINUS

The C-terminus of the ET receptors contains a number of potential sites for palmitoylation which might anchor the receptor to the cell membrane and maintain its conformation (Fig. 3.2). The importance of this region is demonstrated by the fact that removal of the 55 C-terminal amino acids of the human ET_A receptor abolishes the binding of $[^{125}I]$ ET-1.[36] More detailed studies involving sequential truncation of the C-terminus reveal that up to 47 of the 55 amino acids can be removed before there is any substantial effect on ligand binding.[42,51] However, stimulation of increases in intracellular calcium by ET-1 requires the presence of at least 13 C-terminal amino acids, including some potential sites for palmitoylation and for phosphorylation.[42,51]

DISTRIBUTION OF THE ET RECEPTOR GENES

Genes expressing the ET_A and ET_B receptors can be localized by hybridization of the receptor mRNA, either in situ or in total extracted cellular mRNA (Northern blotting), using radiolabeled complementary probes specific for each receptor. These techniques have revealed quite different distributions of the two receptors, although the mechanisms governing tissue specific expression remain elusive. The ET_A receptor is predominantly expressed in vascular smooth muscle, including aorta, coronary vessels and renal arterioles, but also in bronchial smooth muscle, myocardium, adrenal and pituitary gland.[7,19,52] The ET_B receptor is most abundant in vascular endothelial cells and this might account, at least in part, for the substantial levels of ET_B receptor mRNA found in the brain, lung, kidney, stomach, liver, intestine and adrenal glands.[8,11,13,34,52,53]

Within the brain, the highest levels of ET_B receptor mRNA are found in the cerebellum but mRNA can also be detected in the cerebral cortex, brain stem and in glial cells throughout the brain.[11,52] Cheng et al[34] have cloned a novel cDNA from rat brain which they suggest represents an alternative transcript of the ET_B receptor gene and may also exist in other tissues. A very diffuse ET_A receptor signal can be detected over the pituitary,[52] which might correlate with receptor linked inhibition of prolactin release from this tissue by ET-1.[55]

In the human heart, ET_A and ET_B receptor mRNAs have similar distributions; both are found within the atrioventricular node, the penetrating and branching bundles of His, atrial and ventricular myocardium and endocardial cells.[14] The localization of receptor mRNA correlates well with specific binding sites detected in these tissues using autoradiography (Fig. 3.6).[14]

ET$_B$ mRNA has been detected in rat and human kidney[8,11] as has ET$_A$ receptor mRNA, albeit in lower quantity.[7,9] In an elegant study Terada et al[56] localized mRNA encoding each receptor subtype in individual segments of rat kidney using reverse transcription and polymerase chain reaction. Large signals for the ET$_B$ receptor were detected in the initial and terminal inner medullary conducting tubule and outer medullary conducting tubule, vasa recta bundle, and arcuate artery. In contrast, signals for the ET$_A$ receptor were detected only in the glomerulus, vasa recta bundle and arcuate artery. Both ET$_A$ and ET$_B$ mRNA is expressed in rat mesangial cells.[57] Autoradiography using selective ligands for ET$_A$ and ET$_B$ receptors suggests that the ET$_B$ receptor subtype receptor subtype is also dominant in the human kidney, with a particularly high density of ET$_B$-like binding sites distributed in the region of the collecting ducts.[58]

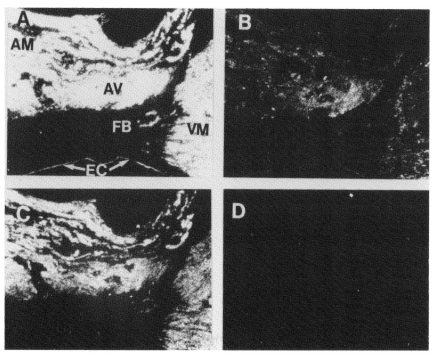

Fig. 3.6. Autoradiography showing binding of [^{125}I]ET-1 to ET$_A$ and ET$_B$ receptors in serial sections of human heart containing atrioventricular node (AV), atrial myocardium (AM), ventricular myocardium (VM), central fibrous body (FB) and endothelial cells (EC). Film images of [^{125}I]ET-1 binding are in the absence (A) and presence of the ET$_A$ selective ligand BQ-123 (B), the ET$_B$ selective ligand, BQ-3020 (C) and unlabeled ET-1 (D). Photograph kindly provided by Dr. A. Davenport and reproduced with permission from Circulation 1993; 72: 526-538.

In vascular tissue, ET_A receptor mRNA is predominant in smooth muscle, while only ET_B mRNA has been detected in most studies of endothelial cells.[10,11,14,59] Some reports have stated that ET_B mRNA cannot be detected in vascular smooth muscle,[8,52] but recently it has been detected both in the medial smooth muscle of arteries[60,61] and in cultured smooth muscle cells.[59,62] These findings are consistent with the evidence for a functional role of ET_B receptors in mediating contraction of vascular smooth muscle (chapter 4). Stanimirovic et al[63] have recently reported detection of ET_A receptor mRNA in cultured cerebro-microvascular endothelial cells. Other studies have failed to detect ET_A receptor mRNA in endothelial cells derived from the peripheral vasculature,[59,60] raising the possibility that endothelial ET_A receptors may be a unique feature of the specialized endothelial cells lining the brain microvasculature.

ALTERATION OF ET RECEPTOR GENE EXPRESSION AND STABILITY

It is well recognized that exposure to ETs for periods in excess of several hours leads to a reduction in the density of ET receptors.[66] Inhibition of endogenous ET production by incubating cultured fibroblasts or mesangial cells with the ECE inhibitor phosphoramidon results in an increased number of binding sites for ET.[67,68] Possible mechanisms to account for a reduction in cell surface receptors include receptor internalization or feedback inhibition of receptor expression.[66] Alternatively, Sakurai et al[69] suggest that endothelial cell ET_B receptor number is more likely to be decreased because exposure to high local concentrations of ET-1 reduces the stability of mRNA molecules, rather than decreasing transcription of receptor mRNA.

Modification of ET receptor expression occurs in several pathological states and may be associated with disease progression (see chapter 7). This can be due to activation of some of the regulatory elements on the ET receptor genes. For example, up-regulation of vascular smooth muscle cell ET_A receptor mRNA by insulin most likely involves an insulin responsive element on the receptor gene promoter, common to genes transcriptionally activated by insulin.[70] In contrast, ET receptor expression is reduced in atherosclerotic human arteries[59] and in the lungs of rats with pulmonary hypertension.[71] In both cases altered receptor expression was accompanied by increased expression of ET-1 mRNA. It cannot be excluded that the change in receptor mRNA may therefore have occurred secondary to the change in local ET-1 concentration rather than to a direct effect on receptor gene transcription. The differential tissue distribution of ET receptors suggests that specific transcription factors regulate the expression of ET receptor genes. The mechanisms governing selective expression might also be implicated in pathological states, allowing one receptor subtype to be expressed preferentially over another. For example, ET_B receptor mRNA is selectively increased in marmosets fed a high cholesterol diet[13] and following glycerol-induced acute renal failure in rats.[72]

FUTURE DIRECTIONS

Despite intensive research over several years since the cloning of the ET-1 selective ET_A receptor[7] and the isopeptide nonselective ET_B receptor,[8] no further subtypes have been identified in the mammalian genome. The question remains as to whether the ET_C[22] and ET_{AX}[23] receptors cloned from the *Xenopus* melanophores and heart are simply amphibian forms of the mammalian receptors or whether they represent separate receptors which have not yet been identified in the mammalian genome. However, functional evidence does exist for receptors with similar characteristics to the ET-3 selective ET_C receptor and BQ-123 insensitive ET_{AX} receptor in mammalian tissue (see chapter 4).

As discussed above, application of molecular biology techniques has already provided clues as to the domains of the receptor that confer binding affinity and ligand selectivity to the ET_A and ET_B receptors. If the current attempts by several groups at purification and crystallization of the ET receptors are successful, then it will be possible to identify more precisely the sites at which receptor and ligand interact. Together, this information should enable the rational design of potent and selective receptor agonists and antagonists.

REFERENCES

1. Martin ER, Brenner BM, Ballermann BJ. Heterogeneity of cell surface endothelin receptors. J Biol Chem 1990; 265: 14044-14049.
2. Jones CR, Hiley CR, Pelton JT et al. Endothelin receptor heterogeneity: structure activity, autoradiographic and functional studies. J Recept Res 1991; 11: 299-310.
3. Emori T, Hirata Y, Marumo F. Specific receptors for endothelin-3 in cultured bovine endothelial cells and its cellular mechanism of action. FEBS Lett 1990; 263: 261-264.
4. Emori T, Hirata Y, Kanno K et al. Endothelin-3 stimulates production of endothelium-derived nitric oxide via phosphoinositide breakdown. Biochem Biophys Res Commun 1991; 174: 228-235.
5. Samson WK, Skala KD, Alexander BD et al. Pituitary site of action of endothelin: selective inhibition of prolactin release in vitro. Biochem Biophys Res Commun 1990; 169: 737-743.
6. Yanagisawa M, Kurihara H, Kimura S et al. A novel potent vasoconstrictor peptide produced by vascular endothelial cells. Nature 1988; 332: 411-415.
7. Arai H, Hori S, Aramori I et al. Cloning and expression of a cDNA encoding an endothelin receptor. Nature 1990; 348: 730-732.
8. Sakurai T, Yanagisawa M, Takuwa Y et al. Cloning of a cDNA encoding a non-isopeptide selective subtype of the endothelin receptor. Nature 1990; 348: 732-735.

9. Lin HY, Kaji EH, Winkel GK et al. Cloning and functional expression of a vascular smooth muscle endothelin 1 receptor. Proc Natl Acad Sci USA 1991; 88: 3185-3189.

10. Hosada K, Nakao K, Arai H et al. Cloning and expression of human endothelin-1 receptor cDNA. FEBS Lett 1991; 287: 223-26.

11. Ogawa Y, Nakao K, Arai H et al. Molecular cloning of a non-isopeptide-selective human endothelin receptor. Biochem Biophys Res Commun 1991; 178: 248-255.

12. Adachi M, Yang Y-Y, Furuichi Y et al. Cloning and characterization of cDNA encoding human A-type endothelin receptor. Biochem Biophys Res Commun 1991; 180: 1265-1272.

13. Elshourbagy NA, Korman DR, Wu HL et al. Molecular characterization and regulation of the human endothelin receptors. J Biol Chem 1993; 268: 3873-3879.

14. Molenaar P, O'Reilly G, Sharkey A et al. Characterization and localization of endothelin receptor subtypes in the human atrioventricular conducting system and myocardium. Circulation 1993; 72: 526-538.

15. Haendler B, Hechler U, Schleuning WD. Molecular cloning of human endothelin (ET) receptors ET_A and ET_B. J Cardiovasc Pharmacol 1992; 20, Suppl 12: S1-S4.

16. Nakamuta M, Takayanagi R, Sakai Y et al. Cloning and sequence analysis of a cDNA encoding human non-selective type of endothelin receptor. Biochem Biophys Res Commun 1991; 177: 34-39.

17. Sakamoto A, Yanagisawa M, Sakurai T et al. Cloning and functional expression of human cDNA for the ET_B endothelin receptor. Biochem Biophys Res Commun 1991; 178: 656-663.

18. Hosada K, Nakao K, Tamura N et al. Organization, structure, chromosomal assignment and expression of the human endothelin-A receptor. J Biol Chem 1992; 267: 18797-18804.

19. Yang H, Tabuchi H, Furuichi Y et al. Molecular characterization of the 5'-flanking region of human genomic ET_A gene. Biochem Biophys Res Commun 1993; 190: 332-339.

20. Arai H, Nakao K, Takaya K et al. The human endothelin B receptor gene: structural organization and chromosomal assignment. J Biol Chem 1993; 268: 3463-3470.

21. Hayzer DJ, Rose PM, Lynch JS et al. Cloning and expression of a human endothelin receptor: subtype A. Am J Med Sci 1992; 304: 231-238.

22. Karne S, Jayawickreme CK, Lerner MR. Cloning and characterization of an endothelin-3 specific receptor (ET_C receptor) from *Xenopus laevis* dermal melanophores. J Biol Chem 1993; 268: 19126-19133.

23. Kumar C, Mwangi V, Nuthalaguni P et al. Cloning and characterization of a novel endothelin receptor from *Xenopus* heart. J Biol Chem 1994; 269: 13414-13420.

24. Elshourbagy NA, Lee JA, Korman DR et al. Molecular cloning and characterization of the major endothelin receptor subtype in porcine cerebellum. Mol Pharmacol 1992; 41: 465-473.

25. Kadonoga JT, Jones KA, Tjan R. Promoter-specific activation of RNA polymerase II transcription by Sp 1. Trends Biochem Sci 1986; 11: 20-23.
26. Wilson D, Dorfman D, Orkin S. A non-erythroid GATA-binding protein is required for function of the human preproendothelin-1 promoter in endothelial cells. Mol Cell Biol 1990; 10: 4854-4862.
27. Lee M-E, Temizer D, Clifford J et al. Cloning of the GATA-binding protein that regulates endothelin-1 gene expresion in endothelial cells. J Biol Chem 1991; 266: 16188-16192.
28. Fowlkes DM, Mullis NT, Corneau CM et al. Potential basis for regulation of the coordinately expressed fibrinogen genes: homology in the 5' flanking regions. Proc Natl Acad Sci USA 1984; 81: 2313-2316.
29. Benoist C, Chambon P. In vivo sequence requirements of the SV40 early promoter region. Nature 1981; 290: 304-310.
30. Mizuno T, Saito Y, Itakura M et al. Structure of the bovine ET_B endothelin receptor gene. Biochem J 1992; 287: 305-309.
31. Burbach JPH, Meijer OC. The structure of neuropeptide receptors. Eur J Pharmacol Mol Pharmacol 1992; 227: 1-18.
32. Trump-Kallmeyer S, Hoflack J, Bruinvels A et al. Modeling of G-protein coupled receptors: application to dopamine, adrenaline, serotonin, acetylcholine, and mammalian opsin receptors. J Med Chem 1992; 35: 3448-3462.
33. Nagayama Y, Wadsworth HL, Chazenbalk GD et al. Thyrotropin-luteinizing hormone/chorionic gonadotropin receptor extracellular domain chimeras as probes for thyrotropin receptor function. Proc Natl Acad Sci USA 1991; 88: 902-905.
34. Cheng HF, Su YM, Yeh JR et al. Alternative transcript of the nonselective-type endothelin receptor from rat brain. Mol Pharmacol 1993; 44: 533-538.
35. Kosuka M, ito T, Hirose S et al. Purification and characterization of bovine lung endothelin receptor. J Biol Chem 1991; 266: 16892-16896.
36. Hashido K, Gamou T, Adachi M et al. Truncation of N-terminal extracellular or C-terminal intracellular domains of human ET_A receptor abrogated the binding activity to ET-1. Biochem Biophys Res Commun 1991; 187: 1241-1248.
37. Takasuka T, Adachi M, Miyamoto C et al. Characterization of endothelin receptors ET_A and ET_B expressed in COS cells. J Biochem 1992; 112: 396-400.
38. Takasuka T, Sakurai T, Goto K et al. Human endothelin receptor ET_B: amino acid requirements for superstable complex formation with its ligand. J Biol Chem 1994; 269: 7509-7513.
39. Strader CD, Sigal IS, Rands E et al. Identification of residues required for binding to the β-adrenergic receptor. Proc Natl Acad Sci USA 1987; 84: 4384-4388.
40. Zhu G, Wu LH, Mauzy C et al. Replacement of lysine-181 by aspartic acid in the third transmembrane region of endothelin type B receptor reduces its affinity to endothelin peptides and sarafotoxin 6c without affecting G-protein coupling. J Cell Biochem 1992; 50: 159-164.

41. Adachi M, Yang YY, Trzeciak A et al. Identification of a domain of ET$_A$ receptor required for ligand binding. FEBS Lett 1992; 311: 179-183.

42. Adachi M, Hashido K, Trzeciak A et al. Functional domains of human endothelin receptor. J Cardiovasc Pharmacol 1993; 22: S121-S124.

43. Ihara M, Noguchi K, Saeki T et al. Biological profiles of highly potent novel endothelin antagonists selective for the ET$_A$ receptor. Life Sci 1992; 50: 247-255.

44. Kubota T, Kamada S, Hirata Y et al. Synthesis and release of endothelin-1 by human decidual cells. J Clin Endocrinol Metab 1992; 75: 1230-1234.

45. Adachi M, Furuichi Y, Miyamoto C. Identification of a ligand binding site of the human endothelin-A receptor and specific regions required for ligand selectivity. Eur J Biochem 1994; 220: 37-43.

46. Sakamoto A, Yanagisawa M, Sawamura T et al. Distinct subdomains of human endothelin receptors determine their selectivity to endothelin A-selective antagonist and endothelin B-selective agonists. J Biol Chem 1993; 268: 8547-8553.

47. Sakamoto A, Yanagisawa M, Sakurai T et al. The ligand-receptor interactions of the endothelin systems are mediated by distinct "message" and "address" domains. J Cardiovasc Pharmacol 1993; 22: 113-116.

48. Krystek SR, Patel PS, Rose PM et al. Mutation of peptide binding site in transmembrane region of a G-protein-coupled receptor accounts for endothelin receptor subtype selectivity. J Biol Chem 1994; 269: 12383-12386.

49. Lee JA, Elliot JD, Sutiphong JA et al. Tyr-129 is important to the peptide ligand affinity and selectivity of human endothelin type A receptor. Proc Natl Acad Sci USA 1994; 91: 7164-7168.

50. Ligett SB, Caron, MG, Lefkowitz RJ et al. Coupling of a mutated form of the human β$_2$ adrenergic receptor to Gi and Gs. J Biol Chem 1991; 266: 4816-4821.

51. Hashido K, Adachi M, Gamou T et al. Identification of specific intracellular domains of the human ET$_A$ receptor required for ligand binding and signal transduction. Cell Mol Biol Res 1993; 39: 3-12.

52. Hori S, Komatsu Y, Shigemoto R et al. Distinct tissue distribution and cellular location of two messenger ribonucleic acids encoding different subtypes of rat endothelin receptors. Endocrinology 1992; 130: 1885-1895.

53. Mizuno T, Imai T, Itakura M et al. Structure of the bovine endothelin-B receptor gene and its tissue- specific expression revealed by Northern analysis. J Cardiovasc Pharmacol 1992; 20, Suppl 12, S34-S37

54. Helfman T, Falanga V. Gene expression in low oxygen tension. Am J Med Sci 1993; 306: 37-41.

55. Samson, WK. The endothelin-A receptor subtype transduces the effect of the endothelins in the anterior pituitary gland. Biochem Biophys Res Commun 1992; 187: 590-595

56. Terada Y, Tomita K, Nonoguchi H et al. Different localization of two types of endothelin receptor mRNA in microdissected rat nephron segments using reverse transcription and polymerase chain reaction assay. J Clin Invest 1992; 90: 107-112.

57. Simonson MS, Rooney A. Characterization of endothelin receptors in mesangial cells: evidence for two functionally distinct endothelin binding sites. Mol Pharmacol 1994; 46: 41-50.

58. Karet FE, Kuc RE, Davenport AP. Novel ligands BQ123 and BQ3020 characterize endothelin receptor subtypes ET_A and ET_B in human kidney. Kidney Int 1993; 44: 36-42.

59. Winkles JA, Alberts GF, Brogi E et al. Endothelin-1 and endothelin receptor mRNA expression in normal and atherosclerotic human arteries. Biochem Biophys Res Commun 1993; 191: 1081-1088.

60. Davenport A, O'Reilly G, Molenaar P et al. Human endothelin receptors characterized using reverse transcriptase-polymerase chain reaction, in situ hybridization, and subtype-selective ligands BQ123 and BQ3020: evidence for expression of ET_B receptors in human vascular smooth muscle. J Cardiovasc Pharmacol 1993; 22: S22-S25.

61. Maguire JJ, Kuc RE, O'Reilly G et al. Vasoconstrictor endothelin receptors characterized in human renal artery and vein. Br J Pharmacol 1994; 113: 49-54.

62. Batra VK, McNeill JR, Xu Y et al. ET_B receptors on aortic smooth muscle cells of spontaneously hypertensive rats. Am J Physiol 1993; 264: C479-C484.

63. Stanimirovic DB, McCarron RM, Spatz M. Dexamethasone down-regulates endothelin receptors in human cerebromicrovascular endothelial cells. Neuropeptides 1994; 26: 145-152.

64. Mikkola T, Ristimaki A, Viinikka L et al. Human serum, plasma and platelets stimulate prostacyclin and endothelin-1 synthesis in human vascular endothelial cells. Life Sci 1993; 53: 283-289.

65. Kohno M, Yokokawa K, Horio T et al. Release mechanism of endothelin-1 and big endothelin-1 after stimulation with thrombin in cultured porcine endothelial cells. J Vasc Res 1992; 29: 56-63.

66. Hirata Y, Yoshimi H, Takaichi S et al. Binding and receptor down-regulation of a novel vasoconstrictor endothelin in cultured rat vascular smooth muscle cells. FEBS Lett 1988; 239: 13-17.

67. WuWong JR, Chiou WJ, Opgenorth T. Phosphoramidon modulates the number of endothelin receptors in cultured Swiss 3T3 fibroblasts. Mol Pharmacol 1993; 44: 422-429.

68. Clozel M, Loffler BM, Breu V et al. Downregulation of endothelin receptors by autocrine production of endothelin-1. Am J Physiol 1993; 265: C188-C192.

69. Sakurai T, Morimoto H, Kasuya Y et al. Level of ET_B receptor mRNA is down-regulated by endothelins through decreasing the intracellular stability of mRNA molecules. Biochem Biophys Res Commun 1992; 186: 342-347.

70. Frank HJL, Levin ER, Hu RM et al. Insulin stimulates endothelin binding and action on cultured vascular smooth muscle cells. Endocrinology 1993; 133: 1092-1097.

71. Yorikane R, Miyauchi T, Sakai S et al. Altered expression of ET_B-receptor mRNA in the lung of rats with pulmonary hypertension. J Cardiovasc Pharmacol 1993; 22, Suppl 8: S336-S338.
72. Roubert P, Gillard-Roubert V, Pourmarin L et al. Endothelin receptor subtypes A and B are up-regulated in an experimental model of acute renal failure. Mol Pharmacol 1994; 182-188.

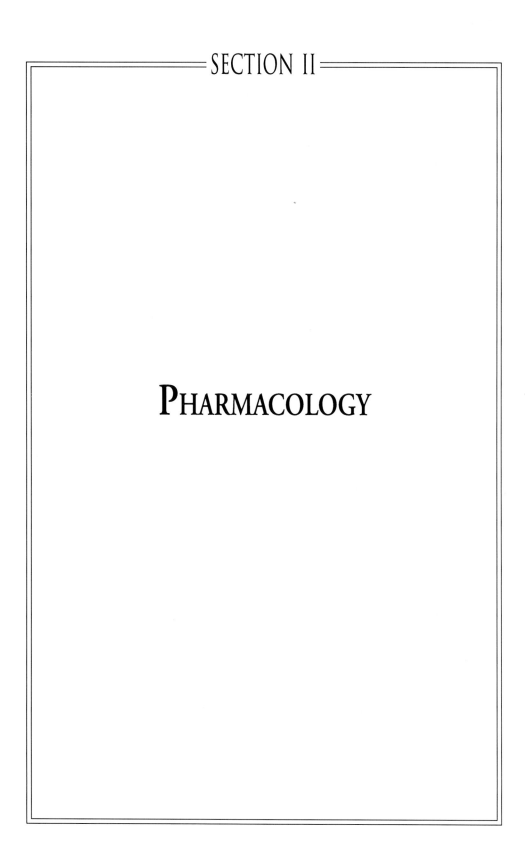

SECTION II

PHARMACOLOGY

CHAPTER 4

PHARMACOLOGICAL CHARACTERIZATION OF ENDOTHELIN RECEPTORS

Gillian A. Gray

Soon after the first description of ET,[1] high affinity binding sites were described in a number of tissues.[2-7] In the majority of these studies, Scatchard analysis suggested that the binding sites were of uniform affinity. However, some of the studies presented evidence for the existence of two distinct cell surface ET binding sites.[4,8] One of these sites bound ET-1 and ET-2, but not ET-3, and the other bound all three isopeptides with equal affinity. Pharmacological studies also pointed to the existence of at least two receptor subtypes mediating the effects of the ETs with orders of agonist potency matching the order of binding affinity.[9,10] For example, in the anesthetized rat, ET-1 was found to be a more potent pressor agent than ET-2 or ET-3 while all three isopeptides were equally effective in evoking vasodilatation.[11] In the previous chapter we described cloning of the ET-1 selective ET_A receptor[12] and the nonisopeptide selective ET_B receptor.[13] The present chapter will consider the nature of the interaction of ligands with the ET_A and ET_B receptor subtypes, including ligand-binding characteristics, receptor internalization and down-regulation. The major part of this chapter will focus on the identification of selective ligands that interact with the ET_A and ET_B subtypes, the structural factors which confer selectivity on the ligands and the development of subtype selective and nonselective receptor antagonists. In the course of these studies, anomalies have arisen which are best explained by the existence of further heterogeneity amongst the ET receptors; the evidence for this is discussed below.

Molecular Biology and Pharmacology of the Endothelins edited by Gillian A. Gray and David J. Webb. © 1995 R.G. Landes Company.

FORMATION AND INTERNALIZATION
OF THE ET/RECEPTOR COMPLEX

In vascular smooth muscle the action of ET-1 can be differentiated from other constrictor agents, not only by its high potency but also by the nature of the contraction it induces. Compared to other vasoconstrictors the action of ET-1 develops more slowly, is more sustained and is more difficult to reverse.[1] In isolated rat aortic rings, full development of tone in response to ET-1 takes up to 40 minutes, compared to approximately 10 minutes or less for other vasoconstrictors like noradrenaline, vasopressin or angiotensin II. The slow onset of contraction is unlikely to be due to the rate of association of ET-1 with its receptors. At a concentration of ET-1 that produces a large increase in tension (50nM), receptor sites are predicted to be saturated with the peptide in less than 1 minute at 37°C.[14] A more likely explanation is the time delay involved in full activation of the second messenger systems within the smooth muscle cell that link occupation of the receptor to contraction. This is discussed further in chapter 5.

Once established, contraction induced by ET is sustained and difficult to wash out, approximately 10% of contraction remains even after 1 hour of washing.[1] While this can also be attributed partially to signaling mechanisms within the cell,[15] the duration and reversibility of the responses is influenced by the essentially irreversible nature of the interaction between ET and its receptors. Binding experiments demonstrate that once ET is bound to the receptor it hardly dissociates, even in the presence of GTP analogues.[2,3,16] The irreversibility of ET action seems to be a function of the exposure time to ET. In rat aortic strips, contraction in response to ET-1 is reversible if exposure time is limited to 30 seconds, but is progressively sustained and irreversible as the exposure time is increased to 15 minutes.[15] This pattern is also seen in binding experiments. In the SK-N-MC neuronal cell line the proportion of nondissociable [^{125}I] ET-1 increases with time over a 60 minute incubation period.[17] Like other vasoconstricting substances, as ET-1 binds to its receptors it is time-dependently internalized within the cell.[3,18] One possible explanation for the results observed in the above studies is that only the receptors remaining on the cell surface bind ET-1 in a reversible manner. As more ET-receptor complex becomes internalized, less of the binding is reversible. Irreversible binding might not be a characteristic shared by all ET isopeptides or all receptors. In cultured rat aortic smooth muscle cells, binding of [^{125}I] ET-1 to ET_A-like binding sites is poorly reversible by unlabeled ET-1, in agreement with the earlier studies.[19] In contrast, binding of [^{125}I] ET-3 to ET_B-like binding sites, in the same cells, is completely displaced by unlabeled ET-1 or ET-3. Both ET_A and ET_B receptors are known to be rapidly internalized.[18,20] Thus, if reversibility is linked in some way to internalization of the receptors, ET-3 must interact differently with ET_B receptors than does ET-1 with ET_A receptors.

Internalization of surface bound [^{125}I] ET-1 is known to occur in a number of cells, including human vascular smooth muscle cells,[18,20] pituitary cells[19] and thymocytes.[21] Surface bound ET-1 can be differentiated from internalized ET-1 by its ability to be dissociated from cells under acid conditions. Using this technique, Hirata et al[20] showed that the ET binding site on cultured rat aortic smooth cells is rapidly internalized following binding of ET-1. The rate and extent of endocytosis of the [^{125}I] ET-receptor complex is dependent on incubation temperature and time. In pituitary cells, acid-resistant binding can be detected within 15 minutes of exposure to [^{125}I]ET-1 at 37°C;[22] while at 4°C, bound ET-1 is completely dissociable even after 2 hour exposure time. Inhibition of endocytosis at low temperatures can be exploited to investigate binding of ET to its receptors without the influence of internalization. For example, at 4°C, ET-1 bound to ET_B receptors on pheochromocytoma cells is more difficult to remove by acid hydrolysis than when it is bound to ET_A receptors on A10 smooth muscle cells or mesangial cells.[8] Perhaps this is related to the reported ability of ET_B, but not ET_A receptors, to form a super-stable complex with ET-1.[23] Internalization of ET-1 by vascular smooth muscle cells is inhibited by dansylcadaverine, showing that ET receptors are internalized by the clathrin-mediated endocytotic pathway.[18] Autoradiographic studies demonstrate the presence of ET-1 binding sites in both the plasma membrane and intracellular compartments of aortic smooth muscle cells.[20] Löffler et al.[24] found that a significant portion of ET-1, -2 and -3 binding sites in the human liver are localized in lysosomes, presumably indicating their participation in the internalization process. The acidic environment in lysosomes would promote dissociation of ET from its receptors, allowing them to be recycled to the cell surface, while at the same time providing suitable conditions for the subsequent breakdown of ET (chapter 2). Internalization of the ET-receptor complex might also be a requirement for the sustained phase of the cellular response to ET-1. In the case of angiotensin II, internalization of the ligand-receptor complex is a prerequisite for sustained diacylglycerol production.[25]

RECEPTOR DOWN-REGULATION

Prolonged exposure of vascular smooth muscle cells to ET-1 leads to a substantial reduction of ET binding sites without any effect on binding affinity.[20] This homologous down-regulation is reflected in functional studies as a loss of responsiveness to ET-1. For example, exposure of cultured neuroblastoma cells[26] or intestinal smooth muscle strips[27] to ET-1 causes either reduction or complete loss of functional responsiveness to further stimulation of the ET_A receptor. Down-regulation is not unique to ET-1 or to ET_A receptors. In guinea-pig ileum, both ET-1 and ET-3 induce homologous desensitisation,[28] while in cultured rat smooth muscle cells ET-1 down-regulates both ET_A and ET_B type binding sites and ET-3 preferentially down-regulates the ET_B type.[19] Down-regulation of ET receptors by endogenously produced ET-1 can be demonstrated in cultured cell systems. For example, the number of binding sites for [^{125}I] ET-1 on cultured Swiss 3T3 fibroblasts[29] or rat mesangial cells[30] is increased following incubation with phosphoramidon to inhibit endogenous ET-1 generation from big ET-1. The ability of ET receptors to be selectively down-regulated can be exploited to investigate the role of different receptor subtypes in mediating functional responses. Exposure of tissues to a high concentration of the selective ET_B receptor agonist, sarafotoxin S6c (SRTX S6c), in vitro, prevents further activation of ET_B receptors by SRTX S6c, or by other ET_B selective ligands, without influencing responses mediated via ET_A receptors.[31,32] In several tissues, including pituitary cells,[22] rat aorta and mesenteric bed,[33] prior exposure to high concentrations of ET-3 prevents further activation of responses by ET-3, but does not affect the ability of ET-1 to evoke a response. This might occur because ET-3 preferentially desensitizes the ET_B receptor, but in at least some of these studies, competitive binding studies fail to detect any ET_B-like binding sites. From this we would have to conclude either that ET-3 desensitizes the ET_A receptor to ET-3 but not ET-1, or alternatively that there is a selective ET-3 receptor to which [^{125}I] ET-1 does not bind. Evidence suggesting the existence of such an ET-3 selective receptor is reviewed below.

It has been suggested that receptor-mediated endocytosis might play a role in the down-regulation process. However, lowering the temperature in pituitary gonadotrophs effectively decreases the extent of receptor internalization but does not affect the development of refractoriness to ET-1 stimulation.[22] Furthermore, homologous desensitization of osteoblastic cells to ET-1 is not inhibited by monodansylcadaverine showing that internalization is most likely not responsible for receptor down-regulation.[34]

Refractoriness to ET-1 can be induced by other agents, in addition to ET-1 itself. This heterologous downregulation is seen with bradykinin in neuroblastoma cells[26] and with both Ang II and phorbol ester in vascular smooth muscle cells.[35] None of these agents binds to the same sites as ET-1. However, a common factor between them is activation of protein kinase C in the target cell, suggesting that this may play a role in the down-regulation mechanism.

STRUCTURAL FEATURES
THAT DETERMINE RECEPTOR SELECTIVITY

In the same year that ET-1 was first described by Yanagisawa and his colleagues,[1] Kloog et al described isolation of the sarafotoxins (SRTXs) from the venom of the Israeli burrowing asp *Atractaspis engaddensis*.[36] As discussed in the first chapter of this book, the SRTXs bear remarkable structural similarity to the ETs and bind competitively to ET receptors. Isoforms of SRTX are utilized extensively as tools in the characterization of ET receptors. Among these isoforms, SRTX S6b binds with high affinity to the ET$_A$ subtype[7] and SRTX S6c is selective for the ET$_B$ subtype (Table 4.1).[37]

Comparison of structures between members of the ET family and between the ET and the SRTX families reveals the features that may be important in determining receptor binding selectivity. Only the major structural requirements can be discussed within the scope of this book, for a more detailed discussion of the structure-activity relationships of the ETs the reader is referred to several excellent review articles.[38-40]

About 60% of the 21 amino acid residues within the ET and SRTX families are identical (chapter 1). Each of the peptides possesses four cysteinyl residues that form two disulphide bridges, three polar charged side chains (residues 8-10) and a well conserved hydrophobic C-terminus (residues 16-21). ET-1 can be considered as a nonselective agonist that exhibits subnanomolar affinities for both receptor subtypes while ET-3 is a moderately ET$_B$ selective agonist and SRTX S6c is completely ET$_B$ receptor selective (up to μM concentration).[37] Most sequence variability amongst these peptides is found in the loop portions, especially within the inner loop (Cys 3-11). The sequence Asp-Lys-Glu at positions 8-10 is well-conserved, except for a substitution of Glu for Lys in SRTX S6c, implicating this region as a particularly important determinant of ET$_B$ selectivity. Variations in the amino terminus might also influence binding activity. In ET-1, ET-2, SRTX S6a and SRTX

Table 4.1 Relative selectivities of agonists at ET$_A$ and ET$_B$ receptors

Agonist	IC$_{50}$/*K$_i$ ET$_A$	Reference	IC$_{50}$/*K$_i$ ET$_B$	Reference	Relative selectivity ET$_A$:ET$_B$
ET-1	3 nM, 1 nM, 0.3 nM	170-172	1 nM, 0.19 nM, 0.2 nM, 1.6 nM	104,172-174	1:1
ET-2	6 nM, 2 nM	170,171	1 nM, 0.18 nM	173,174	1:≈3
ET-3	>6 μM, 0.7 μM, 70 nM	170-172	1 nM, 0.2 nM, 0.2 nM	172-174	1:≈1000
SRTX S6b	0.95 nM	38	0.13 nM	38	1:7
SRTX S6c	>1 μM, 4.5 μM	172,37	0.6 nM, 20pM, 0.03 nM	172,37,176	1:≈300 000
[4-Ala] ET-1	9.9 μM	49	12 nM	49	1:≈1000
IRL 1620	>1 μM*, 1.9 μM*	55,175	0.016 nM*	55	1:>10 000
BQ-3020	940 nM	54	0.2 nM	54	1:4700

S6b, the amino terminal Cys residue is followed by Ser, whereas in ET-3 and SRTX S6c the Cys residue is followed by Thr. Although the two amino acids are structurally related, Bdolah et al[41] postulate that even such a small change may be sufficient to alter the peptide function. Another factor might be the relative charge of the molecules. Sokolovsky et al[38] suggest that expression of ET_A-like activity requires a net charge of -1 within the inner loop, as is the case for ET-1 and ET-2, but not for SRTX S6c (net -4) and ET-3 (net 0).

Chemical modification or truncation of the ET-1 molecule has been used to investigate the structural features important for ligand binding to the ET_A and ET_B receptors. Application of these techniques shows that the ET_A receptor requires both the amino terminal loop portion and the carboxyl terminal linear portion for high affinity binding, and in particular the terminal tryptophan.[42] Reduction of the four cysteinyl residues[42] or replacement of the cysteinyl residues by alanyl residues[43] to produce a linear peptide results in loss of constrictor activity at the ET_A receptor. Similarly, the binding affinity and agonist potency of the SRTXs at the ET_A receptor are lost if non-natural disulphide bonds are formed by connecting Cys1-11 and Cys3-15.[43,44] Interestingly, replacement of only two cysteinyl residues in [Ala1,15] and [Ala3,11] ET-1 does not completely destroy the ability of ET-1 to constrict rat aorta[45] or mesenteric bed[46] suggesting that both disulphide bonds are not essential to maintain the loop structure. The outer disulphide bond (Cys1-Cys15) seems to be more important in maintaining ET-1 activity at the ET_A receptor than the inner disulphide (Cys3-Cys11), since [Ala1,15] ET-1 has a 250 fold decrease in potency relative to the native ET-1, whereas [Ala3,11] ET-1 has only a threefold decrease in activity.[47] [Ala3,11] ET-1 is also more potent than [Ala1,15] ET-1 as a constrictor of the rat aorta[45] and perfused mesenteric bed.[46]

In contrast to the ET_A receptor, the amino terminal loop structure is not an essential requirement for ligands with high affinity at the ET_B receptor. [Ala1,3,11,15] ET-1 is equipotent with ET-1 and ET-3 in inhibiting [125I] ET-1 binding to rat cerebellar homogenate which is rich in ET_B receptors.[16] Furthermore, [Ala1,3,11,15] ET-1 binds to ET_B receptors with an affinity 1700 times greater than that for ET_A receptors, and can elicit ET_B receptor-mediated systemic vasodilatation[48] and relaxation.[49] The linear peptide maintains its activity at endothelial ET_B receptors even when up to nine amino-acid residues are removed from the N-terminal end. However, both high affinity binding and agonist activity are lost following removal of either Glu10 or Trp21.[49,50] Thus, as at the ET_A receptor, the terminal tryptophan moiety seems essential for high affinity binding and activity.[42] Both Glu10 and Trp21 are conserved between all of the ET/SRTX peptides. In early reports, Maggi et al[10,51] described the ability of the C-terminal hexapeptide, ET-1(16-21) to discriminate between various smooth muscle preparations. One set of preparations, typified by the rat aorta, did not respond to the hexapeptide and was termed ET_A

receptor containing, while the others in which ET-1(16-21) was a full agonist, typified by the rat bronchus, were termed ET_B receptor containing. This nomenclature continues to be the one currently recognized by IUPHAR (International Union of Pharmacology nonmenclature of endothelin receptors).[52] Subsequent studies have shown that ET-1(16-21) binds with low affinity to both types of receptors[53] and lacks significant agonist activity,[49] reducing its usefulness as a tool to investigate ET_B receptors. However, ligands with high affinity for ET_B receptors have recently been produced through chemical modification of the carboxyl-terminal region of ET-1. BQ-3020 is a linear ET-1 analogue (N-acetyl- [Ala11,15] ET-1 (6-21)) that has some 4700 fold higher affinity for ET_B than ET_A receptors.[54] Another linear analogue that shows relative selectivity for binding to ET_B receptors is IRL 1620 (N-succinyl-[Glu9, Ala11,15] ET-1 (8-21).[55]

Sakamoto et al[56] have attempted to analyze the structure-activity relationships of ET receptor ligands by applying a message-address concept originally developed by Schwyzer to analyze the structure-activity relationship of ACTH.[57] The principle of this concept is that each ligand has an 'address' component that provides binding affinity and a 'message' component that is responsible for signal transduction. In the case of the ETs, it is proposed that the amino-terminal loop domain of ET-1 functions in a manner similar to the 'address' domain while the carboxyl terminal is the 'message' domain that promotes full activation of the receptor.[56] Based on their results using chimeric ET receptors (chapter 3) and ligands with selectivity at ET_A or ET_B receptors, Sakamoto et al[56] propose that the amino terminal loop portion of ET-1 interacts specifically with transmembrane domains IV-VI of the ET_A receptor. On the other hand, ET_B selective ligands are unable to interact with this region of the ET_A receptor because their address domains are either invalid, in the case of ET-3, SRTX S6c, or missing, in the case of BQ 3020 and IRL 1620. Furthermore, they suggest that these ligands can interact with the ET_B receptor because its subtly different structure within the transmembrane segments must not be so highly specific for the ligand address domains. If this theory were to prove correct, its application is potentially valuable for the design of structures with an intact address domain but no valid message domain that might function as effective ET receptor antagonists.

ENDOTHELIN RECEPTOR ANTAGONISTS

The first compounds with the ability to inhibit the binding or actions of the endothelins were described as early as 1991. Since then, and especially during the past year, details of a large number of compounds with ET receptor antagonist activity have been published. Among these are peptide and nonpeptide compounds, some of which are selective for the ET_A or ET_B subtypes and some that are nonselective. The characteristics of most of these are summarized in Table 4.2 and discussed below. Potent peptide antagonists have been obtained by

Table 4.2. Endothelin receptor antagonists

Selectivity	Compound	Company	Potency	Comments	Reference
ET$_A$ peptide	[Dpr1, Asp15] ET-1	Research Corporation Technology	2 nM*	chemical modification of ET-1	Spinella et al, 1991 Proc Natl Acad Sci USA 88: 7443-7446
	BQ-123	Banyu	7.3 nM*	culture broth of *Streptomyces misakiensis*	Ihara et al, 1991 Life Sci 50: 247-255
	BQ-153			cyclic pentapeptides	Fukoroda et al, 1992 Life Sci 50: PL107-112
	BQ-485	Banyu	3.4 nM*	linear tripeptides	Itoh et al, 1993 Biochem Biophys Res Commun 195: 969-975
	BQ-610		?		Verheyden et al, 1994 FEBS Lett 1994; 344:55-60
	FR-139317	Fujisawa	0.53 nM*	culture broth of *Streptomyces sp.*	Sogabe et al, 1993 J Pharmacol Exp Therap 264: 1040-1046
	TTA-386	Takeda	0.34 nM* Kd = 0.05 nM∞	pseudo-tripeptide synthetic hexapeptide	Kitada et al, 1993 J Cardiovasc Pharmacol 22: Suppl 8, S128-S131
	PD151242	Parke-Davis (Warner-Lambert)	Kd = 0.5-1.7 nM¶	pseudo-tripeptide	Davenport et al, 1994 Br J Pharmacol 111: 46
ET$_A$ non-peptide	50-235	Shionogi	78 nM†	bayberry *Myrica cerifera* myriceron caffeoyl ester	Fujimoto et al, 1992 FEBS Lett 304: 41-44
	97-139	Shionogi	1 nM†	modification of 50-235	Mihara, S et al, 1994 J Pharmacol Exp Therap 288: 1122-
	asterric acid	Kirin Brewery Co	10 µM*	culture broth of *Aspergillus sp.*	Ohashi et al, 1992 J Antibiot 45: 1684-1685
	BMS-182874	Bristol-Myers Squibb	150 nM* 55 nM†	benzenesulfonamide orally active	Stein et al, 1994 J Med Chem 37: 329-331

Selectivity	Compound	Company	Potency	Comments	Reference
ET_B peptide	IRL 1038, [Cys11-Cys15]-ET-1(11-21)	Ciba-Geigy	6-11 nM†	chemical modification of ET-1 (unstable) ET_{B1} selective	Urade et al, 1992 FEBS Lett 311: 12-16
	BQ-788	Banyu	1.2 nM*	tripeptide derivative ET_{B1} and ET_{B2}	Ishikawa et al, 1994 Proc Natl Acad Sci 91: 4892-4896
	RES-701-1	Kyowa Hakko Kogyo Co	10 nM*	culture broth of *Streptomyces sp.* 16aa cyclic peptide ET_{B1} selective	Tanaka et al, 1994 Mol Pharmacol 45: 724-730
ET_A & ET_B peptide	PD 142893	Parke-Davis (Warner-Lambert)	ET_A 15 nM* ET_B 150 nM*	linear hexapeptide analogs of ET-1	Cody et al, 1992 J Med Chem 35: 3301-3303
	PD 145065		ET_A 3.5 nM* ET_B 15 nM*		Cody et al, 1993 Med Chem Res 3: 154-162
	cochinmicins	Merck, Sharpe & Dohme	100-200 nM†	cyclodepsipeptides from culture broth of *Microbispora sp.*	Lam et al, 1992 J Antibiot 45: 1709-1716
	[Thr18, gmethylLeu19]ET-1	Takeda	ET_A 0.7 nM* ET_B 0.3 nM*	chemical modification of ET-1	Shimamoto et al, 1993 J Cardiovasc Pharmacol 22: Suppl 8, S107-S110
	TAK-044	Takeda	ET_A 0.1 nM* ET_B 1.8 nM*	cyclichexapeptide	Kikuchi et al, 1994 Biochem Biophys Res Commun 200: 1708-1712

* IC$_{50}$ for inhibition [125I] ET-1 binding
† K$_i$ in binding assay with [125I]ET-1
°°Kd of [125I] TTA-386 in porcine cardiac ventricular membrane
¶Kd of [125I] PD151242 in human aorta, pulmonary and epicardial arteries

Continued...

Table 4.2. Endothelin receptor antagonists *(continued)*

Selectivity	Compound	Company	Potency	Comments	Reference
ET$_A$ & ET$_B$ non-peptide	Ro 46-2005	Hoffmann-La Roche	ET$_A$ 360 nM* ET$_B$ 530 nM*	sulphonamides orally active	Clozel et al, 1993 Nature 365: 759-761
	Ro 47-0203 (bosentan)		ET$_A$ 4.7 nM† ET$_B$ 95 nM†		Clozel et al, 1994 J Pharmacol Exp Therap 270: 228-235
	CGS 27830	Ciba-Geigy	ET$_A$ 16 nM* 5.6 nM† ET$_B$ 295 nM*	irreversible binding, unstable	Mugrage et al, 1993 Bio Med Chem Letts 3: 2099-2104
	SB 209670	SmithKline Beecham	ET$_A$ 0.2 nM† ET$_B$ 18 nM†	orally active	Ohlstein et al, 1994 Proc Natl Acad Sci USA 91: 8052-8051
undefined	JKC-1		na	homperidinyl substituted quartapeptides	Ristea et al, 1994 FASEB J 8: Abs 598
	IPI-413	Immuno Pharmaceutic	low nM*	orally active	Pharma projects
	sulfonic acid polymers		7 nM (rat brain)		Verma et al, 1994 FASEB J 8: Abs 596

* IC$_{50}$ for inhibition [125I] ET-1 binding
† K$_i$ in binding assay with [125I]ET-1
na not available

chemical modification of ET-1 itself, or by using microbial products with ET receptor binding activity. Although useful as research tools, the potential of peptides as therapeutic agents may be limited by their lack of oral availability. Short duration of action is another possible disadvantage of these compounds, a problem that has been encountered with some, but not all, of the recently developed peptide ET antagonists. To circumvent these problems much effort has been applied to the development of nonpeptide antagonists with oral availability and long duration of action. Several potent antagonists have been developed through optimization of compounds isolated from plant extracts and microbial broths, or screened from chemical libraries. Among these, Ro 46-2005 was the first nonpeptide antagonist shown to be orally available,[58] although several other compounds of this sort are now in development.

ET$_A$ Receptor Selective Antagonists

The peptides [Dpr1, Asp15] ET-1[59] and BE-18257B[60] were among the first ET receptor antagonists to be described. These are representative of a series of peptide antagonists prepared by chemical modification of ET-1, like [Dpr1, Asp15] ET-1, or isolated by screening of microbial broth, like BE-18257B. Both of these antagonists preferentially inhibit binding of [^{125}I] ET-1 to the ET-1 selective ET$_A$ receptor. In the case of [Dpr1, Asp15] ET-1, substitution of the outer disulphide bridge of ET-1 for an amide bond causes little change in molecular structure but complete loss of agonist activity.[59] BE-18257B is a cyclic pentapeptide, cyclo(-D-Glu-L-Ala-allo-D-Ile-L-Leu-D-Trp-), that is highly ET$_A$ selective, but lacking in potency relative to ET-1 (IC$_{50}$ of 0.5µM for inhibition of [^{125}I] ET-1 binding to aortic smooth muscle). BQ-123 is a related cyclic pentapeptide, cyclo (D-Trp-D-Asp-Pro-D-Val-Leu-), that maintains the ET$_A$ selectivity of these compounds but has increased affinity and antagonist potency.[61] BQ-123 has become a reference compound for ET$_A$ receptor antagonists, and its use first confirmed the role of ET-1 in a number of pathologies.[62] Although it is said to be a competitive antagonist at ET receptors,[63] there is evidence that BQ-123 may have a noncompetitive interaction with ET$_A$ receptors in some tissues.[64,65] BQ-153 (cyclo-D-sulphalanine-L-Pro-D-Val-L-Leu-D-Trp-) is the most recent in this series of cyclic pentapeptides.[66,67] Knowledge gained from examining the structure activity relationships of the cyclic pentapeptides was used to design a new class of linear tripeptide ET$_A$ selective antagonists, including BQ-485 (perhydroazepin-1-yl-L-leucyl-D-tryptophanyl-D-tryptophan)[68] and BQ-610 (homopiperidinyl-CO-Leu-D-Trp(CHO)-D-Trp-OH).[69,70] These compounds retain several 1000 fold higher affinities for binding to the ET$_A$ than the ET$_B$ receptor, while being much more feasible to synthesize in large quantities than the cyclic pentapeptides. Another group of ET$_A$ selective antagonists derived from microbial products

are pseudo-tripeptides. One of these, FR 139317 has similar antagonist potency at the ET_A receptor to BQ-123.[71] PD 151242 is a synthetic antagonist with a structure related to that of FR 139317. Although it has lower antagonist potency in functional studies than
FR 139317, its ability to be easily radiolabeled, unlike either BQ-123
or FR 139317, makes it a useful tool for the study of ET_A receptors.[72]
The addition of several hydrophobic moieties to another tripeptide
molecule, TTA-101 (Bu^tOCO-Leu-Trp-Ala) led to development of the
novel hexapeptide ET_A receptor antagonist, TTA-386 (hexamethyl-
eneiminocarbonyl-Leu-Trp-Ala-β Ala-Tyr-Phe),[73] that is among the most
potent described to date (Table 4.2).

A naturally occurring caffeoyl ester extracted from the bayberry
Myrica cerifiera was the first nonpeptide ET_A receptor antagonist to
be described.[74] This compound, 50-235, is selective for the ET_A receptor, but low in potency relative to BQ-123. The chemically modified derivative of 50-235, 97-139 (27-O-3-[2[(3-carboxy-acryloylamino)-
5-hydroxyphenyl]-acryloy-loxy myricerone),[75] has considerably improved
affinity for the ET_A receptor (Table 4.2) and a potency greater than
BQ-123 in in vitro functional tests (pA_2 of 8.8 vs. 7.3 for BQ-123).
Asterric acid, a compound derived from the culture broth of a fungus,
Aspergillus sp.[76] has the property of inhibiting ET-1 binding to the
ET_A receptor but its usefulness is limited by its very low potency. In
contrast to these naturally occurring compounds, BMS-182874
(5-(dimethylamino)-N-(3,4-dimethyl-5-isoxazyloyl)-1-napthenesulfon-
amide) is an ET_A selective antagonist produced by optimization of compounds obtained by random screening of a chemical library.[77] Interestingly, this compound is a sulphonamide, as is the nonsubtype selective
antagonist, Ro 46-2005.[58] BMS-182874 is orally active and has a long
duration of action relative to the peptide ET_A antagonists, although
its IC_{50} for inhibition of [^{125}I]ET-1 binding is some 20 fold less than
that of BQ-123 (Table 4.2).

ET_B Receptor Selective Antagonists

Only peptide antagonists selective for the ET_B receptor have
been described to date. The first of these is IRL 1038 ([Cys11,
Cys15]-ET-1(11-21,[78]), designed on the basis that the C-terminal half
of ET-1 is well conserved among the ET isopeptides and thus might
be important for binding to the isopeptide unselective ET_B receptor.
As discussed above, this strategy was previously used to design a selective agonist of ET_B receptors, IRL 1620.[55] IRL 1038 has been used in
a number of studies to identify the presence or role of ET_B receptors.[78-81] However, it has recently been recognized that, although IRL
1038 is relatively potent, it is inherently unstable, complicating interpretation of results obtained with it. An equally potent but more stable
ET_B selective antagonist is the recently described 16 amino acid cyclic
peptide of microbial origin, RES-701-1.[82,83] This compound shares no

amino acid sequence homology with ET-1 except for the carboxyl-terminal tryptophan, reported to be important for ET-1 activity.[42] However, like ET-1, it has a cyclic structure and contains a hydrophobic region, features which may be important for binding to ET receptors.[84] In contrast to ET-1, RES-701-1 contains no charged amino acids, which Tanaka et al[82] postulate may explain its lack of agonist activity relative to ET-1. BQ-788 (N-cis-2,6-dimethylpiperidinocarbonyl-L-gamma-methylleucyl-D-1-methoxycarbonyltryptophanyl-D-norleucine) is another recently described ET_B receptor selective antagonist, with a tripeptide structure related to that of the ET_A selective antagonist FR 139317.[85-87] BQ-788 has an affinity approximately 1000 fold higher for ET_B than for ET_A receptors and a pA_2 of approximately 8 for inhibition of ET_B receptor-mediated dilatation and constriction.[87]

ANTAGONISTS AT BOTH ET_A AND ET_B RECEPTOR SUBTYPES

Despite the reported association of the C-terminal of ET with selectivity for the ET_B receptor, chemical modification of the C-terminal hexapeptide structure [ET-1 (15-21)] leads to compounds with antagonist activity at both ET_A and ET_B receptors.[88,89] The C-terminal hexapeptide itself exhibits weak binding inhibitory potency. However, potency can be enhanced by acetylation of the N-terminal free amine and by substitution of the His residue at position 16, particularly when the substituted amino acid has D-stereochemistry.[89] Two particularly potent nonselective ET receptor antagonists, PD 142893 (Ac-D-Dip16-Leu-Asp-Ile-Ile-Trp21)[90] and PD 145065 (Ac-D-Bhg16-Leu-Asp-Ile-Ile-Trp21) were designed on this basis.[91] Both of these peptides are competitive antagonists with higher potency at ET_A than ET_B receptors. High D-amino acid content, in addition to a number of rare amino acid residues, is a feature shared by the cochinmicins, a group of cyclodepsipeptide compounds isolated by microbial screening of culture broth from *Microbiospora sp* ATCC55140.[92] These compounds bind to both ET_A and ET_B receptor subtypes, although even the most potent of them is 10 to 100 fold less potent than other more recently described antagonists. [Thr18, gamma methyl Leu19] ET-1 is a potent antagonist prepared by chemical modification of the entire ET molecule, that has approximately equal activity at ET_A and ET_B receptor subtypes.[93] The cyclic hexapeptide, TAK-044 has lower affinity for ET_B than for ET_A receptors but behaves as a nonselective antagonist in vivo.[94-97] Long lasting inhibition of both ET_A and ET_B receptor-mediated blood pressure responses by TAK-044 demonstrates that short duration of action is not a problem that besets all peptide antagonists.

Ro 46-2005 was the first of a now growing group of ET receptor antagonists to be designed by optimization of nonpeptide compounds derived from existing chemical libraries. It is a sulphonamide derivative with approximately equal antagonist activity at ET_A and ET_B receptors.[58,98,99] Bosentan (Ro 47-0203; 4-tert-butyl-N-[6-(2-hydroxy-ethoxy)-

5-(2-methoxy-phenoxy0-2,2'-bipyrimidin-4-yl]-benzenesulfonamide) is a structurally optimized derivative of Ro 46-2005 with higher relative selectivity for the ET_A receptor subtype, but enhanced potency at both subtypes.[100] Both of these compounds have competitive antagonist characteristics and are orally active. CGS 27830 is an asymmetric anhydride derived from the dihydropyridine class of calcium channel modulators. This compound has poor potency as an inhibitor of ligand binding to L-type calcium channels but potently inhibits binding of [^{125}I] ET-1 to the ET_A receptor.[101] It is only approximately 18 fold less potent as an inhibitor of ET-1 binding to the ET_B receptor subtype and thus can be considered as a relatively unselective compound. Functional studies in vitro suggest that CGS 27830 is a noncompetitive antagonist of ET receptors. The most recent additions to the non subtype selective group nonpeptide ET antagonists is SB 209670 ([(+)-(1S, 2R, 3S)-3-(2-carboxymethoxy-4-methoxyphenyl)-1-(3,4-methylenedioxyphenyl)-5-(prop-1-yloxy)indane-2-carboxylic acid].[102,103] This compound was rationally designed using conformational models of ET-1.[103] It is more potent at the ET_A than the ET_B receptors in binding and functional studies and is a competitive antagonist at both subtypes.[102]

EVIDENCE FOR FURTHER SUBTYPES OF ET RECEPTOR

Low stringency Southern blot analysis revealed only two ET receptor genes in human genomic DNA, most likely corresponding to the cloned ET_A and ET_B receptors.[104] Genes encoding other receptor subtypes, if they exist, must therefore have quite low sequence homogeneity. However, this does not necessarily rule out further receptor heterogeneity. Bax and Saxena point out in a recent review[32] that amino acid homology between receptors does not necessarily correlate with pharmacological similarity. The cloned 5-HT receptors provide a good example; the $5-HT_{1B}$ and $5-HT_{1D\beta}$ receptors have 96% sequence homology but have quite different pharmacological profiles. Novel ET receptors could result from differential splicing following transcription or by post-translational modification of the expressed receptor proteins. As discussed in the previous chapter, a change of only a few amino acids in the receptor can radically change its binding profile. There is already evidence for alternative RNA splicing of the ET_A receptor gene.[105]

Typical ET_A and ET_B receptor subtypes can be defined by their differing relative affinities for the ET and SRTX isopeptides (Table 4.1 and chapter 3). In general, ET-1 binds with subnanomolar affinity to both subtypes while ET-3 binds with similar affinity to the ET_B subtype but 100-300 fold lower affinity to the ET_A subtype. SRTX S6b is relatively selective for the ET_A receptor subtype while SRTX S6c has 300,000 fold higher affinity for the ET_B than the ET_A receptor. These differences are reflected in the rank order of potency of the peptides in functional studies. A typical ET_A receptor-mediated response is contraction of rat aortic smooth muscle (Fig. 4.1). In this tissue, ET-1 is a more potent constrictor than ET-3; SRTX S6b is active but SRTX S6c is not and the ET-1 dose-response curve is shifted to the right by BQ-123. Endothelium-dependent relaxation of the perfused rat mesenteric bed is typical of an ET_B receptor-mediated response (Fig. 4.1). ET-1, ET-3 and SRTX S6c are equipotent in induction of relaxation and responses to these peptides are unaffected by BQ-123.

Early indications of ET binding sites with characteristics that differed from this classical profile came from competitive binding studies, showing that radiolabeled ETs and SRTXs bind to more than two sites with differing affinities. For example, studies carried out using various tissues of the rat and dog suggest the existence of at least three[7] or even four[106] ET-1 binding sites. Included in these binding sites are two which conform to the ET_A and ET_B receptor subtypes but others with, for example, particularly high affinity for SRTX S6c or for ET-3.[7] Other studies show the existence of binding sites with the same rank order of ligand affinity but quite different relative affinities, perhaps indicative of different subtypes of ET_A and ET_B receptor.[107]

Receptor heterogeneity might provide an explanation for the presence of multiple bands on SDS gel electrophoresis of membranes chemically cross-linked with [^{125}I]ET-1, giving apparent molecular weights of 30 to 70kD for ET binding proteins.[5,7,8,108,109] The predicted molecular weights of the cloned ET receptor cDNAs are in the range of 45-50kD.[12,13] However, caution must be applied to interpretation of cross-linking data since some of the fragments might represent different glycosylation states of the mature receptors,[110] or degradation products arising from proteolytic cleavage.[111]

Evidence that challenges the simple classification of ET receptors into two subtypes comes from several sources: comparison of binding affinities in competitive binding studies; classical pharmacology comparing rank orders of agonist potency and the actions of recently developed selective ET receptor agonists and antagonists; and studies of cell signaling mechanisms linked to receptor activation in various tissues. Some of this literature is reviewed below. For further analysis the reader is referred to several recent reviews dealing with the subject of ET receptor heterogeneity.[107,112]

PUTATIVE ET_A RECEPTOR SUBTYPES

Kumar et al[113] have identified a binding site for ET-1 on follicular membranes of *Xenopus laevis* oocytes that has a typically ET_A type binding profile for the ET isopeptides, but no affinity for the ET_A selective antagonist, BQ-123. As discussed in the previous chapter, this so called ET_{AX} receptor has since been cloned from a *Xenopus* heart cDNA library.[114] The high overall homology of this receptor with the mammalian ET_A receptor and the fact that this receptor has not yet been

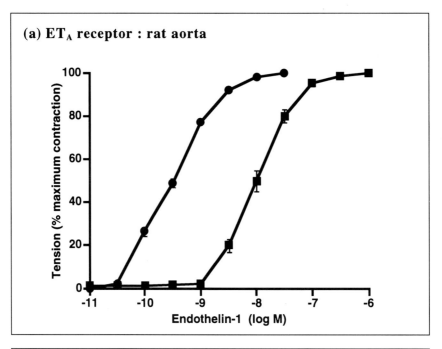

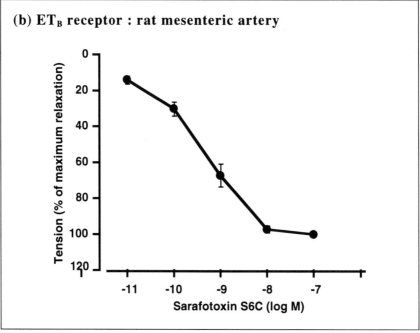

Fig. 4.1. Prototypical ET_A and ET_B receptor-mediated responses in blood vessels. (a) Contraction of the endothelium-denuded rat aorta by ET-1 (-●-) and competitive inhibition of contraction by the ET_A receptor antagonist, BQ-123 ($10^{-6}M$, -■-); (b) Endothelium-dependent relaxation of rat third generation mesenteric arteries by the selective ET_B receptor agonist, sarafotoxin S6c.

identified in the mammalian genome raise the possibility that it might represent an amphibian form of the mammalian receptor. However, several recent reports of BQ-123 insensitive effects of ET-1 at 'typically ET_A' receptors in mammalian tissues might be explained by the existence of such an ET_{AX} receptor. Human umbilical[115] and rat renal arteries[116] constrict in response to ET-1 but not ET-3 or SRTX consistent with ET_A receptor-mediated contraction, but ET-1 evoked contractions are not sensitive to BQ-123. Sudjarwo et al[117] recently proposed the existence of both BQ-123-sensitive (ET_{A1}) and BQ-123-insensitive (ET_{A2}) receptors in the rabbit saphenous vein. This is based on the observation that BQ-123 is a more potent inhibitor of ET-3 than of ET-1 evoked constriction, even after desensitization of the ET_B type receptors that mediate contraction in this preparation. Several other recent studies report a similar disparity in the ability of BQ-123 to inhibit ET-1 and ET-3 induced responses.[118-121] For example, in the rat vas deferens BQ-123 (1µM) inhibits potentiation of the electrically induced twitch response by ET-3 but not by ET-1.[121,122] These results have been interpreted as suggesting that ET-1 and ET-3 bind to different subtypes of the ET_A receptor in some tissues. However, when interpreting results obtained using BQ-123, it should be borne in mind that its affinity for binding to the ET_A receptor is less than that of ET-1, but more than that of ET-3[62] and that it interaction with the ET_A like binding sites may be more complex than simple competitive inhibition.[123] Formal verification of a second ET_A receptor subtype awaits either identification by cloning or the development of other antagonists that can discriminate between subtypes.

ET_A receptors are in general associated with mediation of vascular smooth muscle contraction, while ET_B receptors are linked to relaxation via liberation of relaxing factors from the endothelium or epithelium. However, recent studies, using the ET_A receptor antagonist BQ-123, indicate that ET_A receptors might mediate release of relaxant factors from the endothelium of aorta from hypertensive rats[124] and from the epithelium of guinea-pig trachea.[125] Furthermore, high affinity binding sites typical of ET_A receptors are located on rat brain microvascular endothelial cells where they are linked to activation of phospholipase C.[126] Studies with BQ-123 and the ET_B selective agonist, SRTX S6c, suggest that ET_A type receptors also mediate inhibition of smooth muscle contraction by ET-1 in the rat stomach.[127] To date ET_A receptor mRNA has not been located on either endothelial or epithelial cells, future studies will show whether the receptors present in these cells and in the rat stomach represent a novel class of ET_A receptors.

PUTATIVE ET_B RECEPTOR SUBTYPES

Early reports that ET-1 was a more potent constrictor of smooth muscle than ET-3 lead to the concept that smooth muscle cell contraction was mediated solely by ET_A receptors. Competitive inhibition of contraction by BQ-123 in a number of tissues is consistent with this.[62] However, in some tissues ET-3 and other ET analogues seemed to discriminate between two types of receptor mediating contraction,[128] and competition binding studies demonstrated biphasic inhibition of [^{125}I]ET-1 binding by ET-3.[19] Soon after the description of the first ET_A selective antagonists it became clear that contraction of vascular smooth muscle by ET-1 consisted of both ET_A antagonist sensitive and insensitive components (Fig. 4.2),[67,128,129] and that pressor responses to ET-1 in vivo were not solely due to stimulation of ET_A receptors (chapter 6).[130,131] Subsequent studies using ET_B selective agonists in vitro[118,132,133] and in vivo[37,48,134] showed that ET_B receptors have a vasoconstrictor role. Recently, in situ hybridization has confirmed the presence of ET_B receptor mRNA in the smooth muscle layer of blood vessels.[135,136] The relative contributions of ET_A and ET_B receptors to vasoconstriction is variable and depends on species and the vessel type studied,[137] with indications that ET_B receptors are generally more prevalent in the low pressure venous circulation than in the arterial circulation.[116] In the human blood vessels that have been studied to date, ET_B receptors seem to play a relatively minor role in comparison to ET_A receptors.[119,136-141] However, it is interesting to note that ET_B receptor expression is increased under certain conditions; for example during the change of cultured vascular smooth muscle cells from a contractile to a synthetic phenotype,[142] in hypertension,[143] and under the influence of Ang II.[144] These observations raise the possibility that the role of ET_B receptors might be altered in pathophysiological situations.

Resistance of ET-1 induced contractions to BQ-123 and sensitivity to ET_B selective agonists, indicate that ET_B receptors also mediate contraction in nonvascular smooth muscle preparations; for example in the rat trachea (Fig. 4.2),[31] guinea pig bronchus,[145] human bronchus,[146] guinea-pig ileum,[28,147] and rat stomach.[127,148] Furthermore, comparison of the structure-activity relationships of ET-1 and a number of analogues provides evidence that the ET_B receptors mediating contraction of the guinea-pig trachea might be different from those mediating vascular smooth muscle contraction.[149]

ET_A receptors stimulate contraction via activation of phospholipase C and subsequent activation of protein kinase C(PKC) (see chapter 5). ET-3 also stimulates liberation of nitric oxide via endothelial ET_B receptors linked to activation of phospholipase C and elevation of cytosolic Ca^{2+} levels.[150,151] In contrast, the ET_B receptor mediating contraction of the rabbit saphenous vein does not seem to require PKC activation.[152] This might be because the ET_B receptors couple to different signaling mechanisms in different cell types or, alternatively, that the ET_B receptors described on smooth muscle are different from those on endothelial cells. The latter argument is strengthened by recent data obtained with newly developed ET_B receptor selective antagonists. Both IRL-1038 and RES 701-1 are reported to inhibit endothelium-dependent relaxation while having no effect on smooth muscle constriction evoked by ET_B selective agonists.[79,81,153] The nonselective antagonists, bosentan[100] and PD 142893[148] also have differential potency on the endothelial (so-called ET_{B1}) type of receptor, compared to the constrictor (so-called ET_{B2}) type of receptor (Fig. 4.3). In contrast, the ET_B selective antagonist, BQ-788 has a pA_2 of approximately 8.0

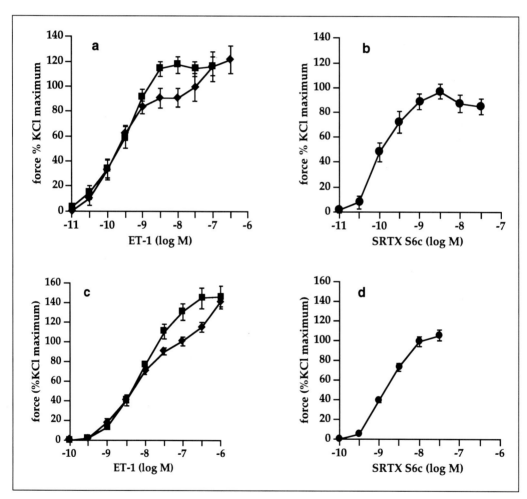

Fig. 4.2. ET receptors mediating contraction of vascular smooth muscle: rabbit saphenous vein (a and b) and nonvascular smooth muscle: rat trachea (c and d). In both tissues ET-1 induced contraction (a and c, -■-) has a component that is resistant and a component that is sensitive to inhibition by the ET_A receptor selective antagonist, BQ-123 ($10^{-6}M$, -◆-). Contraction induced by the selective ET_B receptor agonist sarafotoxin S6c (b and d, -●-) corresponds to the BQ-123 resistant portion of the contraction induced by ET-1. Figs. a and b are reproduced from Am J Physiol 1994; 266: H959-H966 with permission of the American Physiological Society.

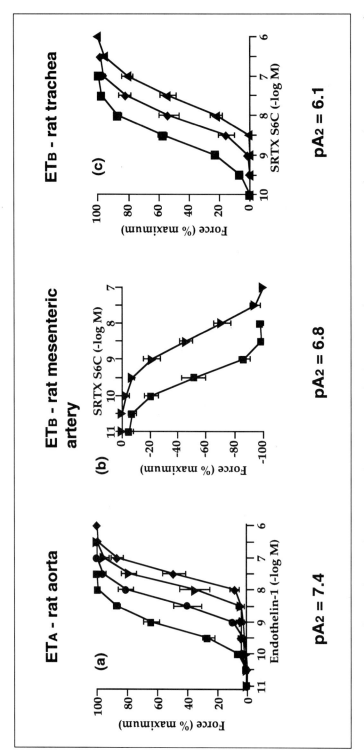

Fig. 4.3. Competitive inhibition of ET-1 and SRTX S6c mediated responses by the nonpeptide ET receptor antagonist bosentan (Ro 47-0203; ■,:control; ●, 0.3 μM; -▼, 1 μM; -◆, 3 μM; -▲, 10 μM). The antagonist pA₂ values calculated for inhibition of (a) ET_A receptor-mediated contraction of the rat aorta; (b) ET_B receptor-mediated endothelium-dependent relaxation of the 3rd generation rat mesenteric artery and (c) ET_B receptor-mediated contraction of the rat trachea suggest that the ET_B receptors might represent separate ET_{B1} and ET_{B2} receptor subtypes (see text). Figure reproduced with permission from Clozel et al, J Pharmacol Exp Ther 1994; 270: 228-235.

for inhibition of both types of response.[87] This could be because the ET_{B1} and ET_{B2} receptors have only minor differences in their structure. Indeed, molecular studies suggest that this might be the case, since probes designed on the basis of the cloned endothelial ET_B receptor are able to detect receptor expression in both endothelial and smooth muscle cells.[135,136]

Competitive binding studies in rabbit saphenous vein and canine coronary artery indicate that smooth muscle may contain more than one ET_B-like binding site.[152,154] The first has nM affinity for ET-1 and SRTX S6c but low capacity and seems to correspond to the receptor mediating contraction in these vessels. A second site with equal but low (μM) affinity for all isopeptides and a high capacity is also identifiable in these vessels. Similar low affinity binding sites for SRTX S6c and [Ala 1,3,11,15] ET-1 are present in rat atria, although neither peptide elicits a response in this tissue.[155] These sites might therefore be linked to functions other than contraction. A role in internalization[156] or clearance of excess concentrations of ET[157] would be more consistent with their low affinity and high capacity. Radioligand binding studies in rat brain and atrium have also identified two ET-1 binding sites with binding profiles typical of ET_B receptors. The first, 'super high affinity' site has picomolar affinity for the ET isopeptides and is not linked to phosphoinositide hydrolysis whereas the second, 'high affinity' site has nanomolar affinity and is linked to phosphoinositide hydrolysis.[158] The former site might correspond to a recently documented ET_B receptor, present on human umbilical vein endothelial cells, that has picomolar affinity and is linked to liberation of NO via a tyrosine-kinase dependent pathway.[159] The latter site might correspond to a receptor on cultured human endothelial cells that is linked to phospholipase C and appears to be selective for ET-3.[160]

Further subtypes of ET_B receptor with different agonist selectivity might exist. In the rat vas deferens and rat atria, ET-1 and ET-3 are equipotent, typical of responses mediated by ET_B receptors, but in neither case is the response reproduced by SRTX S6c.[122,155] A binding site with similar characteristics, termed the ET_{BX} receptor, was recently identified in membranes prepared from *Xenopus laevis* liver.[161] In contrast to this, competitive binding experiments using membranes prepared from rat cerebellum indicate the presence of a high affinity receptor that binds SRTX S6c but not ET-1, ET-3 or SRTX S6b.[7]

PUTATIVE ET_C RECEPTOR

In the previous chapter we described cloning of an ET-3 selective ET_C receptor from *Xenopus laevis* melanophores.[162] The existence of a receptor with these characteristics in mammalian tissues remains controversial. Emori et al[163] report that cultured bovine endothelial cells express a receptor through which ET-3, but not ET-1, evokes increases in intracellular calcium and liberates NO.[150,163] In a separate report it is suggested that an ET-3 selective receptor on these cells mediates prolonged release of EDRF/NO.[164] Cultured human endothelial cells are reported to contain both 'classical' ET_B receptors and ET-3 preferring receptors that are linked to NO release.[160] These results are, however, disputed by others who suggest that ET-3 preferring receptors on endothelial cells are an artifact created by prolonged periods of cell culture.[165] This argument cannot be applied to all tissues that are suggested to contain ET-3 preferring binding sites such as cardiac membranes prepared from newborn chicks.[5] The existence of selective receptors for ET-3 is supported by its high potency relative to ET-1 in a number of functional studies. For example, ET-3 is more a potent agonist than ET-1 in rat renal papillary tubules[166] and in the rat liver, ET-3 induces a greater reduction in tissue blood flow than ET-1.[167] Initial reports suggested that ET-3, but not ET-1, was able to inhibit prolactin secretion in pituitary lactotrophs.[168] More recent studies show that this effect is prevented by BQ-123, suggesting that it is more likely to be an ET_A receptor-mediated event.[169]

FUTURE DIRECTIONS

The above discussion gives some idea of the complexity emerging among the receptors mediating responses to the ET isopeptides. Some of the implied heterogeneity might have alternative explanations, bearing in mind the problems involved in interpretation of data related to internalization and recycling of receptors and the complex allosteric interactions that might occur between some ligands. Even so, it is clear that the current classification of the receptors as ET_A and ET_B may not describe all the existing receptor subtypes. Development of new and selective ligands should help to improve our understanding of the receptor subtypes that exist, but only molecular cloning and further investigation into the post-transcriptional processes that might influence receptor structure will provide a final answer. Obviously, clear delineation of receptor subtypes has important implications for the design of receptor antagonists for therapeutic use. With many of the nonpeptide antagonists already in the first phases of clinical trials it will be interesting to see if they fulfill the early promise they show in experimental studies.

REFERENCES

1. Yanagisawa M, Kurihara H, Kimura S et al. A novel potent vasoconstrictor peptide produced by vascular endothelial cells. Nature 1988; 332: 411-415.
2. Clozel M, Fischli W, Guilly C. Specific binding of endothelin on human vascular smooth muscle cells in culture. J Clin Invest 1989; 83: 1758-1761.
3. Hirata Y, Yoshimi H, Takata S et al. Cellular mechanism of action of a novel vasoconstrictor endothelin in cultured rat vascular smooth muscle cells. Biochem Biophys Res Commun 1988; 154: 868-875.

4. Badr KF, Munger KA, Sugiara M et al. High and low affinity binding sites for endothelin on cultured rat glomerular mesangial cells. Biochem Biophys Res Commun 1989; 161: 776-781.

5. Watanabe H, Miyazaki H, Kondoh M et al. Two distinct types of ET receptors are present on chick cardiac membranes. Biochem Biophys Res Commun 1989; 161: 1252-1259.

6. Cozza EN, Gomez-Sanchez CE, Foecking MF et al. Endothelin binding to cultured calf adrenal zona glomerulosa cells and stimulation of aldosterone secretion. J Clin Invest 1989; 84: 1032-1035.

7. Kloog Y, Bousso-Mittler B, Bdolah A et al. Three apparent receptor subtypes for the endothelin/sarafotoxin family. FEBS Lett 1989; 253: 199-202.

8. Martin ER, Brenner BM, Ballermann BJ. Heterogeneity of cell surface endothelin receptors. J Biol Chem 1990; 265: 14044-14049.

9. Warner TD, De Nucci G, Vane JR. Rat endothelin is a vasodilator in the isolated perfused mesentery of the rat. Eur J Pharmacol 1989; 159: 325-326.

10. Maggi CA, Giuliani S, Patacchini R et al. The activity of peptides of the endothelin family in various mammalian smooth muscle preparations. Eur J Pharmacol 1989; 174: 23-31.

11. Spokes RA, Ghatei MA, Bloom SR. Studies with endothelin 3 and endothelin 1 on rat blood pressure and isolated tissues: evidence for multiple endothelin receptor subtypes. J Cardiovasc Pharmacol 1989; 13, Suppl 5: S191-S192.

12. Arai H, Hori S, Aramori I et al. Cloning and expression of a cDNA encoding an endothelin receptor. Nature 1990; 348: 730-732.

13. Sakurai T, Yanagisawa M, Takuwa Y et al. Cloning of a cDNA encoding a non-isopeptide selective subtype of the endothelin receptor. Nature 1990; 348: 732-735.

14. Marsault R, Vigne P, Breittmayer JP et al. Kinetics of vasoconstrictor actions of endothelins. Am J Physiol 1991; 261: C986-C993.

15. Marsault R, Vigne P, Frelin C. The irreversibilty of endothelin action is a property of a late intracellular signalling event. Biochem Biophys Res Commun 1991; 179: 1408-1413.

16. Hiley CR, Jones CR, Pelton JT et al. Binding of [^{125}I] endothelin-1 to rat cerebellar homogenates and its interactions with some analogues. Br J Pharmacol 1990; 101: 319-324.

17. Wilkes LC, Boarder MR. Characterization of endothelin receptors on a human neuroblastoma cell line: evidence for the ET_A subtype. Br J Pharmacol 1991; 104: 750-754.

18. Resink TJ, Scott-Burden T, Boulanger C et al. Internalization of endothelin by cultured human vascular smooth muscle cells: characterization and physiological significance. Mol Pharmacol 1990; 38: 244-252.

19. Roubert P, Gillard V, Plas P et al. Binding characteristics of endothelin isoforms (ET-1, ET-2 and ET-3) in vascular smooth muscle cells. J Cardiovasc Pharmacol 1991; 17, Suppl 5: S104-S108.

20. Hirata Y, Yoshimi H, Takaichi S et al. Binding and receptor down-regulation of a novel vasoconstrictor endothelin in cultured rat vascular smooth muscle cells. FEBS Lett 1988; 239: 13-17.

21. Jackson S, Tseng Y-C, Lahiri S et al. Receptors for endothelin in cultured human thyroid cells and inhibition by endothelin of thyroglobulin secretion. J Clin Endocrinol Metab 1992; 75: 388-392.

22. Stojilkovic SS, Balla T, Fukuda S et al. Endothelin ET_A receptors mediate the signaling and secretory actions of endothelins in pituitary gonadotrophs. Endocrinology 1992; 130: 465-474.

23. Takasuka T, Sakurai T, Goto K et al. Human endothelin receptor ET_B: amino acid requirements for superstable complex formation with its ligand. J Biol Chem 1994; 269: 7509-7513.

24. Löffler BM, Kalina B, Kunze H. Partial characterization and subcellular distribution patterns of endothelin-1, -2 and -3 binding sites in human liver. Biochem Biophys Res Commun 1991; 181: 840-845.

25. Greindling KK, Delafontaine P, Rittenhouse SE et al. Correlation of receptor sequestration with sustained diacylglycerol accumulation in angiotensin II-stimulated cultured vascular smooth muscle cells. J Biol Chem 1987; 262: 14555-14562.

26. Chau LY, Lin TA, Chang WT et al. Endothelin-mediated calcium response and inositol 1,4,5-trisphosphate release in neuroblastoma-glioma hybrid cells (NG108-15): cross talk with ATP and bradykinin. J Neurochem 1993; 60: 454-460.

27. Bolger GT, Liard F, Garneau M et al. Characterization of intestinal smooth muscle responses and binding sites for endothelin. Can J Physiol Pharmacol 1992; 70: 377-84.

28. Miasiro N, Paiva ACM. Effects of endothelin-1 and endothelin-3 on the isolated guinea pig ileum: role of Na^+ ions and endothelin receptor subtypes. Eur J Pharmacol 1992; 214: 133-141.

29. WuWong JR, Chiou WJ, Opgenorth T. Phosphoramidon modulates the number of endothelin receptors in cultured Swiss 3T3 fibroblasts. Mol Pharmacol 1993; 44: 422-429.

30. Clozel M, Loffler BM, Breu V et al. Downregulation of endothelin receptors by autocrine production of endothelin-1. Am J Physiol 1993; 265: C188-C192.

31. Henry PJ. Endothelin-1(ET-1)-induced contraction in rat isolated trachea: involvement of ET_A and ET_B receptors and multiple signal transduction systems. Br J Pharmacol 1993; 110: 435-441.

32. Bax WA, Bos E, Saxena PR. Heterogeneity of endothelin/sarafotoxin receptors mediating contraction of the human isolated saphenous vein. Eur J Pharmacol 1993; 239: 267-268.

33. Hiley CR, McStay MKG, Bottrill FE et al. Cross-desensitisation studies with endothelin isopeptides in the rat isolated superior mesenteric arterial bed. J Vasc Res 1992; 29: 135.

34. Tatrai A, Stern PH. Endothelin-1 modulates calcium signaling by epidermal growth factor, alpha- thrombin, and prostaglandin E_1 in UMR-106

osteoblastic cells. J Bone Miner Res 1993; 8: 943-952.

35. Roubert P, Gillard V, Plas P et al. Angiotensin II and phorbol ester potently down-regulate endothelin-1 (ET-1) binding sites in vascular smooth muscle cells. Biochem Biophys Res Commun 1989; 164: 809-815.

36. Kloog Y, Ambar I, Sokolovky M et al. Sarafotoxin, a novel vasoconstrictor peptide: phosphoinositide hydrolysis in rat heart and brain. Science 1988; 242: 268-270.

37. Williams DL, Jones KL, Pettibone DJ et al. Sarafotoxin S6c: an agonist which distinguishes between endothelin receptor subtypes. Biochem Biophys Res Commun 1991; 175: 556-561.

38. Sokolovsky M. Structure-function relationships of endothelins, sarafotoxins, and their receptor subtypes. J Neurochem 1992; 59: 809-821.

39. Erhardt PW. Endothelin structure and structure-activity relationships. In: Rubanyi GM, ed(s). Endothelin. New York: Oxford University Press, 1992: 41-47.

40. Huggins JP, Pelton JT, Miller RC. The structure and specificity of endothelin receptors: their importance in physiology and medicine. Pharmacol Therap 1993; 59: 55-123.

41. Bdolah A, Wollberg Z, Fleminger G et al. SRTX d, a new native peptide of the endothelin sarafotoxin family. FEBS Lett 1989; 256: 1-9.

42. Kimura S, Kasuya T, Sawamura O et al. Structure-activity relationships of endothelin: importance of the C-terminal moiety. Biochem Biophys Res Commun 1988; 156: 1182-1186.

43. Hirata Y, Yoshini H, Emori T et al. Receptor binding activity and cytosolic free calcium response by synthetic endothelin analogs in cultured rat vascular smooth muscle cells. Biochem Biophys Res Commun 1989; 160: 228-234.

44. Kitazumi K, Shiba T, Nishiki K et al. Structure-activity relationship in vasoconstrictor effects of sarafotoxins and endothelins. FEBS Lett 1990; 260: 269-272.

45. Topouzis S, Huggins JP, Pelton JT et al. Modulation by endothelium of the responses induced by endothelin-1 and by some of its analogues in rat isolated aorta. Br J Pharmacol 1991; 102: 545-549.

46. Randall MD, Douglas SA, Hiley CR. Vascular activities of endothelin-1 and alanyl substituted analogues in resistance beds of the rat. Br J Pharmacol 1989; 98: 685-699.

47. Nakajima K, Kubo S, Nishio H et al. Structure-activity relationship in endothelin: importance of charged groups. Biochem Biophys Res Commun 1989; 163: 424-429.

48. Bigaud M, Pelton JT. Discrimination between ET_A- and ET_B-receptor-mediated effects of endothelin-1 and (Ala1,3,11,15) endothelin-1 by BQ-123 in the anesthetized rat. Br J Pharmacol 1992; 107: 912-918.

49. Saeki T, Ihara M, Fukuroda T et al. [Ala1,3,11,15] Endothelin-1 analogs with ET_B agonistic activity. Biochem Biophys Res Commun 1991; 179: 286-292.

50. Filep JG, Rousseau A, Fournier A et al. Structure-activity relationship of analogues of endothelin-1: dissociation of hypotensive and pressor actions. Eur J Pharmacol 1992; 220: 263-266.

51. Maggi CA, Giuliani R, Patacchini P et al. The C-terminal hexapeptide, endothelin-(16-21) discriminates between different endothelin receptors. Eur J Pharmacol 1989; 166: 121-125.

52. Masaki T, Vane JR, Vanhoutte PM. International Union of Pharmacology nomenclature of endothelin receptors. Pharmacol Rev 1994; 46: 137-142.

53. Doherty AM. Endothelin: a new challenge. J Med Chem 1992; 35: 1493-1508.

54. Ihara M, Saeki T, Fukuroda T et al. A novel radioligand [^{125}I]BQ-3020 selective for endothelin ET_B receptors. Life Sci 1992; 51: 47-52.

55. Takai M, Unemura I, Yamasaki K et al. A potent and specific agonist, Suc-[Glu9,Ala11,15]-endothelin-1(8-21), IRL 1620, for the ET_B receptor. Biochem Biophys Res Commun 1992; 184: 953-959.

56. Sakamoto A, Yanagisawa M, Sawamura T et al. Distinct subdomains of human endothelin receptors determine their selectivity to endothelin A-selective antagonist and endothelin B-selective agonists. J Biol Chem 1993; 268: 8547-8553.

57. Schwyzer R. ACTH: a short review. Ann N Y Acad Sci 1977; 247: 3-26.

58. Clozel M, Breu V, Burri K et al. Pathophysiological role of endothelin as revealed by the first orally active endothelin receptor antagonist. Nature 1993; 365: 759-761.

59. Spinella MJ, Malik AB, Everitt J et al. Design and synthesis of a specific endothelin 1 antagonist: effects on pulmonary vasoconstriction. Proc Natl Acad Sci USA 1991; 88: 7443-7446.

60. Ihara M, Fukuroda T, Saeki T et al. An endothelin receptor ET_A antagonist isolated from *Streptomyces misakiensis*. Biochem Biophys Res Commun 1991; 178: 132-137.

61. Ihara M, Noguchi K, Saeki T et al. Biological profiles of highly potent novel endothelin antagonists selective for the ET_A receptor. Life Sci 1992; 50: 247-255.

62. Moreland S. BQ-123, a selective endothelin ET_A receptor antagonist. Cardiovasc Drug Rev 1994; 12: 48-69.

63. Eguchi S, Hirata Y, Ihara M et al. A novel ET_A antagonist (BQ-123) inhibits endothelin-1-induced phosphoinositide breakdown and DNA synthesis in rat vascular smooth muscle cells. FEBS Lett 1992; 302: 243-246.

64. Hiley CR, Cowley DJ, Pelton JT et al. BQ-123, cyclo(-C-Trp-D-Asp-Pro-D-Val-Leu), is a non-competitive antagonist of endothelin-1 in SK-N-MC human neuroblastoma cells. Biochem Biophys Res Commun 1992; 184: 504-510.

65. Vigne P, Breittmayer JP, Frelin C. Competitive and non-competitive interactions of BQ-123 with endothelin ET_A receptors. Eur J Pharmacol-Mol Pharmacol Section 1993; 245: 229-232.

66. Cirino M, Motz C, Maw J et al. BQ-153, a novel endothelin ET_A an-

tagonist, attenuates the renal vascular effects of endothelin-1. J Pharm Pharmacol 1992; 44: 782-785.

67. Fukuroda T, Nishikibe M, Ohta Y et al. Analysis of responses to endothelins in isolated porcine blood vessels by using a novel endothelin antagonist, BQ-153. Life Sci 1992; 50: PL107-PL112.

68. Itoh S, Sasaki T, Ide K et al. A novel endothelin ET_A receptor antagonist, BQ-485, and its preventive effect on experimental cerebral vasospasm in dogs. Biochem Biophys Res Commun 1993; 195: 969-975.

69. Verheyden P, Assche IV, Brichard MH et al. Conformational study of three endothelin antagonists with ^1H NMR at low temperature and molecular dynamics. FEBS Lett 1994; 344: 55-60.

70. Zuccarello M, Lewis AI, Rapoport RM. Endothelin ET_A and ET_B receptors in subarachnoid hemorrhage-induced cerebral vasospasm. Eur J Pharmacol 1994; 259: R1-R2.

71. Sogabe K, Nirei H, Shoubo M et al. Pharmacological profile of FR139317, a novel, potent endothelin ET_A receptor antagonist. J Pharmacol Exp Ther 1993; 264: 1040-1046.

72. Davenport A, Kuc R, Fitzgerald F et al. [^{125}I] PD15142: a selective radioligand for human ET_A receptors. Br J Pharmacol 1994; 111: 4-6.

73. Kitada C, Ohtaki T, Masuda Y et al. Design and synthesis of ET_A receptor antagonists and study of ET_A receptor distribution. J Cardiovasc Pharmacol 1993; 22, Suppl 8: S128-S131.

74. Fujimoto M, Mihara S, Nakajima S et al. A novel non-peptide endothelin antagonist isolated from bayberry, *Myrica cerifera*. FEBS Lett 1992; 305: 41-44.

75. Mihara S-I, Nakajima S, Matumura S et al. Pharmacological characterization of a potent nonpeptide endothelin receptor antagonist, 97-139. J Pharmacol Exp Ther 1994; 268: 1122-1128.

76. Ohashi H, Akiyama H, Nishikori K et al. Asterric acid, a new endothelin binding inhibitor. J Antibiotics 1992; 45: 1684-1685.

77. Stein PD, Hunt JT, Floyd DM et al. The discovery of sulfonamide endothelin antagonists and the development of the orally active ET_A antagonist 5-(dimethylamino) -N-(3,4-dimethyl-5- isoxazolyl)-1-naphthalenesulfonamide. J Med Chem 1994; 37: 329-331.

78. Urade Y, Fujitani Y, Oda K et al. An endothelin B receptor-selective antagonist: IRL 1038, (Cys11- Cys15)-endothelin-1(11-21). FEBS Lett 1992; 311: 12-16.

79. Karaki H, Sudjarwo SA, Hori M et al. ET_B receptor antagonist, IRL 1038, selectively inhibits the endothelin-induced endothelium-dependent vascular relaxation. Eur J Pharmacol 1993; 231: 371-374.

80. Sudjarwo SA, Hori M, Karaki H. Effect of endothelin-3 on cytosolic calcium level in vascular endothelium and on smooth muscle contraction. Eur J Pharmacol 1992; 229: 137-142.

81. Sudjarwo SA, Hori M, Takai M et al. A novel subtype of endothelin B receptor mediating contraction in swine pulmonary vein. Life Sci 1993; 53: 431-437.

82. Tanaka T, Tsukuda E, Nozawa M et al. RES-701-1, a novel, potent, endothelin type B receptor-selective antagonist of microbial origin. Mol Pharmacol 1994; 45: 724-730.

83. Morishita Y, Chiba S, Tsukuda E et al. RES-701-1, a novel and selective endothelin type B receptor antagonist produced by *Streptomyces sp*. RES-701. I. Characterization of producing strain, fermentation, isolation, physico-chemical and biological properties. J Antibiotics 1994; 47: 269-275.

84. Kloog K, Sokolovsky M. Similarities in the mode and sites of action of endothelins and sarafotoxins. Trends Pharmacol Sci 1989; 10: 212-214.

85. Ishikawa K, Ihara M, Noguchi K et al. Biochemical and pharmacological profile of a potent and selective endothelin B-receptor antagonist, BQ-788. Proc Natl Acad Sci USA 1994; 91: 4892-4896.

86. Fukuroda T, Ozaki S, Ihara M et al. Synergistic inhibition by BQ-123 and BQ-788 of endothelin-1-induced contractions of the rabbit pulmonary artery. Br J Pharmacol 1994; 113: 336-338.

87. Fujikawa T, Fukuroda T, Yano M et al. The sensitivity of endothelin ET_B receptors on vascular endothelium and smooth muscle to a novel ET_B antagonist, BQ-788. Jpn J Pharmacol 1994; 64: Abstract P-433.

88. Doherty AM, Cody WL, Leitz NL et al. Structure-activity studies of the C-terminal region of the endothelins and the sarafotoxins. J Cardiovasc Pharmacol 1991; 17, Suppl. 7: S59-S61.

89. Doherty AM, Cody WL, He JX et al. Design of C-terminal peptide antagonists of endothelin. Structure-activity relationships of ET-1(16-21, D-His16). Bioorg Med Chem Lett 1993; 3: 497-502.

90. Cody WL, Doherty AM, He JX et al. Design of a functional hexapeptide antagonist of endothelin (1). J Med Chem 1992; 35: 3301-3303.

91. Cody WL, Doherty AM, He JX et al. The rational design of a highly potent combined ET_A and ET_B receptor antagonist (PD 145065) and related analogues. Med Chem Res 1993; 3: 154-162.

92. Lam KT, Williams DL, Sigmund JM et al. Cochinmicins, novel and potent cyclodepsipeptide endothelin antagonists from a *Microbiospora* sp. 1. Production, isolation and characterization. J Antibiotics 1992; 45: 1709-1716.

93. Shimamoto N, Kubo K, Watanabe T et al. Pharmacologic profile of endothelin A/B antagonist, [Thr18, gmethylLeu19] endothelin-1. J Cardiovasc Pharmacol 1993; 22, Suppl. 8: S107-S110.

94. Kikuchi T, Ohtaki T, Kawata A et al. Cyclic hexapeptide endothelin receptor antagonists highly potent for both receptor subtypes ET_A and ET_B. Biochem Biophys Res Commun 1994; 200: 1708-1712.

95. Ikeda S, Awane Y, Kusumoto K et al. A new endothelin receptor antagonist, TAK-044, shows long-lasting inhibition of both ET_A- and ET_B-mediated blood pressure responses in rats. J Pharmacol Exp Ther 1994; 270: 728-733.

96. Awane Y, Kusumoto K, Kubo K et al. Pharmacological profile in vivo of a new endothelin receptor antagonist, TAK-044. Jpn J Pharmacol 1994; 64, suppl 1: 167P.

97. Kusumoto K, Kubo K, Kandori H et al. Effects of a new endothelin antagonist, TAK-044, on post-ischemic acute renal failure in rats. Life Sci 1994; 55: 301-310.

98. Breu V, Löffler BM, Clozel M. In vitro characterization of Ro 46-2005, a novel synthetic non-peptide endothelin antagonist of ET_A and ET_B receptors. FEBS Lett 1993; 334: 210-214.

99. Clozel M, Breu V, Gray GA et al. In vivo pharmacology of Ro 46-2005, the first synthetic nonpeptide endothelin receptor antagonist: implications for endothelin pathophysiology. J Cardiovasc Pharmacol 1993; 22, Suppl 8: S377-S379.

100. Clozel M, Breu V, Gray GA et al. Pharmacological characterization of bosentan, a new orally active non-peptide endothelin receptor antagonist. J Pharmacol Exp Ther 1994; 270: 228-235.

101. Mugrage B, Moliterni J, Robinson L et al. CGS 27830, a potent nonpeptide endothelin receptor antagonist. FEBS Lett 1993; 3: 2099-2104.

102. Ohlstein EH, Nambi P, Douglas SA et al. SB 209670, a rationally designed potent nonpeptide endothelin receptor antagonist. Proc Natl Acad Sci USA 1994; 91: 8052-8056.

103. Elliot JD, Lago JD, Cousins RD et al. 1,3-diarylindan-2-carboxylic acids, potent and selective non-peptide endothelin receptor antagonists. J Med Chem 1994; 37: 1553-1557.

104. Sakamoto A, Yanagisawa M, Sakurai T et al. Cloning and functional expression of human cDNA for the ET_B endothelin receptor. Biochem Biophys Res Commun 1991; 178: 656-663.

105. Miyamoto Y, Yoshimasa T, Arai H et al. Alternative RNA splicing of the human endothelin-A receptor gene. Circulation 1994; 90: Abstract 386.

106. Löffler B-M, Löhrer W. Different endothelin receptor affinities in dog tissues. J Recept Res 1991; 11: 293-298.

107. Sokolovsky M. Endothelins and sarafotoxins: receptor heterogeneity. Internat J Biochem 1994; 26: 335-340.

108. Schvartz I, Ittoop O, Hazum E. Identification of endothelin receptors by chemical cross-linking. Endocrinology 1990; 126: 1829-1833.

109. Sugira M, Snadjar RM, Schwartzberg M et al. Identification of two types of specific endothelin receptors in rat mesangial cell. Biochem Biophys Res Commun 1989; 162: 1396-1401.

110. Bousso-Mittler D, Galron R, Sokolovsky M. Endothelin/sarafotoxin receptor heterogeneity: evidence for different glycosylation in receptors from different tissues. Biochem Biophys Res Commun 1991; 178: 921-926.

111. Saito Y, Mizuno T, Itakura M et al. Primary structure of bovine endothelin ET_B receptor and identification of signal peptidase and metal proteinase cleavage sites. J Biol Chem 1991; 266: 23433-23437.

112. Bax WA, Saxena PR. The current endothelin receptor classification: time for reconsideration? Trends Pharmacol Sci 1994; 15: 379-386.

113. Kumar CS, Nuthulaganti P, Pullen M et al. Novel endothelin receptors in the follicular membranes of *Xenopus laevis* oocytes mediate calcium responses by signal transduction through gap junctions. Mol Pharmacol 1993; 44: 153-157.

114. Kumar C, Mwangi V, Nuthalaguni P et al. Cloning and characterization of a novel endothelin receptor from *Xenopus* heart. J Biol Chem 1994; 269: 13414-13420.

115. Bodelsson G, Sternquist M. Characterization of endothelin receptors and localization of ^{125}I-endothelin-1 binding sites in human umbilical artery. Eur J Pharmacol 1993; 249: 299-305.

116. Moreland S D, Abboa-Offei B, Seymour A. Evidence for a differential location of vasoconstrictor endothelin receptors in the vasculature. Br J Pharmacol 1994; 112: 704-708.

117. Sudjarwo SA, Hori M, Tanaka T et al. Subtypes of endothelin ET_A and ET_B receptors mediating venous smooth muscle contraction. Biochem Biophys Res Commun 1994; 200: 627-633.

118. Sumner MJ, Cannon TR, Mundin JW et al. Endothelin ET_A and ET_B receptors mediate vascular smooth muscle contraction. Br J Pharmacol 1992; 107: 858-860.

119. Riezebos J, Watts IS, Vallance PT. Endothelin receptors mediating functional responses in human small arteries and veins. Br J Pharmacol 1994; 111: 609-615.

120. Godfraind T. Evidence for heterogeneity of endothelin receptor distribution in human coronary artery. Br J Pharmacol 1993; 110: 1201-1205.

121. Warner TD, Allcock GH, Mickley EJ et al. Characterization of endothelin receptors mediating the effects of the endothelin/sarafotoxin peptides on autonomic neurotransmission in the rat vas deferens and guinea-pig ileum. Br J Pharmacol 1993; 110: 783-789.

122. Eglezos A, Cucchi P, Patacchini R et al. Differential effects of BQ-123 against endothelin-1 and endothelin-3 on the rat vas deferens: evidence for an atypical endothelin receptor. Br J Pharmacol 1993; 109: 736-738.

123. Sokolovsky M. BQ-123 identifies heterogeneity and allosteric interactions at the rat heart endothelin receptor. Biochem Biophys Res Commun 1993; 196: 32-38.

124. Taddei S, Vanhoutte PM. Role of endothelium in endothelin-evoked contractions in the rat aorta. Hypertension 1993; 21: 9-15.

125. Battistini B, Warner TD, Fournier A et al. Characterization of ET_B receptors mediating contractions induced by endothelin-1 or IRL 1620 in guinea-pig isolated airways: effects of BQ-123, FR 139317 or PD 145065. Br J Pharmacol 1994; 111: 1009-1116.

126. Frelin C, Ladoux A, Marsault R et al. Functional properties of high- and low-affinity receptor subtypes for ET-3. J Cardiovasc Pharmacol 1991; 17, Suppl. 7: S131-S133.

127. Gray GA, Clozel M. ET-1 activates multiple receptor subtypes, including an inhibitory ET_A receptor in the rat stomach strip. Br J Pharmacol 1994;

128. Ihara M, Saeki K, Funabashi K et al. Two endothelin receptor subtypes in porcine arteries. J Cardiovasc Pharmacol 1991; 17, Suppl. 7: S119-S121.

129. Schoeffter P, Randriantsoa A, Jost B et al. Comparative effects of the two endothelin ET_A receptor antagonists, BQ-123 and FR139317, on

endothelin-1-induced contraction in guinea-pig iliac artery. Eur J Pharmacol 1993; 241: 165-169.

130. McMurdo L, Corder R, Thiemermann C et al. Incomplete inhibition of the pressor effects of endothelin-1 and related peptides in the anaesthetized rat with BQ-123 provides evidence for more than one vasoconstrictor receptor. Br J Pharmacol 1993; 108: 557-561.

131. Cristol JP, Warner TD, Thiemermann C et al. Mediation via different receptors of the vasoconstrictor effects of endothelins and sarafotoxins in the systemic circulation and renal vasculature of the anaesthetized rat. Br J Pharmacol 1993; 108: 776-779.

132. Moreland S, McMullen DM, Delaney CL et al. Venous smooth muscle contains vasoconstrictor ET_B-like receptors. Biochem Biophys Res Commun 1992; 184: 100-106.

133. Shetty SS, Okada T, Webb RL et al. Functionally distinct endothelin ET_B receptors in vascular endothelium and smooth muscle. Biochem Biophys Res Commun 1993; 191: 459-464.

134. Clozel M, Gray GA, Breu V et al. The endothelin ET_B receptor mediates both vasodilation and vasoconstriction in vivo. Biochem Biophys Res Commun 1992; 186: 867-873.

135. Winkles JA, Alberts GF, Brogi E et al. Endothelin-1 and endothelin receptor mRNA expression in normal and atherosclerotic human arteries. Biochem Biophys Res Commun 1993; 191: 1081-1088.

136. Davenport AP, O'Reilly G, Molenaar P et al. Human endothelin receptors characterized using reverse transcriptase-polymerase chain reaction, in-situ hybridization and subtype selective ligands BQ123 and BQ3020: evidence for expression of ET_B receptors in human vascular smooth muscle. J Cardiovasc Pharmacol 1993; 22, Suppl 8: S22-S25

137. Davenport AP, Maguire JJ. Is endothelin-induced vasoconstriction mediated only by ET_A receptors in humans? Trends Pharmacol. Sci. 1994; 15: 9-11.

138. Maguire J, Kuc RE, O'Reilly G et al. Vasoconstrictor endothelin receptors characterised in human renal artery and vein in vitro. Br J Pharmacol 1994; 113: 49-54.

139. Opgaard OS, Adner M, Gulbenkan S et al. Localisation of endothelin immunoreactivity and demonstration of vasoconstrictory endothelin-A receptors in human coronary arteries and veins. J Cardiovasc Pharmacol 1994; 23: 576-583.

140. Wenzel RR, Noll G, Luscher TF. Endothelin receptor antagonists inhibit endothelin in human skin microcirculation. Hypertension 1994; 23: 581-586.

141. White DG, Garratt H, Mundin JW et al. Human saphenous vein contains both endothelin ET_A and ET_B contractile receptors. Eur J Pharmacol 1994; 257: 307-310.

142. Eguchi S, Hirata Y, Imai T et al. Phenotypic change of endothelin receptor subtype in cultured rat vascular smooth muscle cells. Endocrinology 1994; 134: 222-228.

143. Batra VK, McNeill JR, Xu Y et al. ET_B receptors on aortic smooth muscle cells of spontaneously hypertensive rats. Am J Physiol 1993; 264: C479-C484.

144. Kanno K, Hirata Y, Tsujino M et al. Upregulation of ET_B receptor subtype mRNA by angiotensin II in rat cardiomyocytes. Biochem Biophys Res Commun 1993; 194: 1282-1287.

145. Hay DWP. Pharmacological evidence for distinct endothelin receptors in guinea pig bronchus and aorta. Br J Pharmacol 1992; 106: 759-761.

146. Hay DW, Hubbard WC, Undem BJ. Endothelin-induced contraction and mediator release in human bronchus. Br J Pharmacol 1993; 110: 392-8.

147. Yoshinoga M, Chijiiwa Y, Misawa T et al. Endothelin B receptor on guinea-pig small intestinal smooth muscle cells. Am J Physiol 1992; 262: G308-G311.

148. Warner TD, Allcock GH, Corder R et al. Use of the endothelin antagonists BQ-123 and PD 142893 to reveal three endothelin receptors mediating smooth muscle contraction and the release of EDRF. Br J Pharmacol 1993; 110: 777-782.

149. Battistini B, Germain M, Fournier A et al. Structure-activity relationships of ET-1 and selected analogues in the isolated guinea-pig trachea: evidence for the existence of diferent ET_B receptor subtypes. J Cardiovasc Pharmacol 1993; 22, Suppl 8.: S219-S224.

150. Emori T, Hirata Y, Kanno K et al. Endothelin-3 stimulates production of endothelium-derived nitric oxide via phosphoinositide breakdown. Biochem Biophys Res Commun 1991; 174: 228-235.

151. Hirata Y, Emori T, Eguchi S et al. Endothelin receptor subtype B mediates synthesis of nitric oxide by cultured bovine endothelial cells. J Clin Invest 1993; 91: 1367-1373.

152. Gray GA, Löffler BM, Clozel M. Characterization of endothelin receptors mediating contraction of the rabbit saphenous vein. Am J Physiol 1994; 266: H959-H966.

153. Karaki H, Sudjarwo SA, Hori M et al. Effects of a novel endothelin ET_B receptor antagonist, RES-701-1, on isolated blood vessels. Jpn J Pharmacol 1994; 64: Abstract O-338.

154. Teerlink JR, Breu V, Clozel C et al. Potent vasoconstriction mediated by endothelin ET_B receptors in canine coronary arteries. Circulation 1994; 74: 105-114.

155. Panek RL, Major TC, Hingoranm GP et al. Endothelin and structurally related analogues distinguish between endothelin receptor subtypes. Biochem Biophys Res Commun 1992; 184: 566-571.

156. Gandhi CR, Behal RH, Harvey SAK et al. Hepatic effects of endothelin. Biochem J 1992; 287: 897-904.

157. Marsault R, Vigne P, Breittmayer J-P et al. High reactivity of aortic fibroblasts to vasoactive agents: endothelins, bradykinin and nucleotides. Biochem Biophys Res Commun 1992; 188: 205-208.

158. Sokolovsky M, Ambar I, Galron R. A novel subtype of endothelin receptors. J Biol Chem 1992; 267: 20551-20554.

159. Tsukuhara H, Ende H, Magazine HI et al. Molecular and functional characterization of the non-isopeptide-selective ET$_B$ receptor in endothelial cells. J Biol Chem 1994; 269: 21778-21785.

160. Yokokawa K, Kohno M, Yasunuri K et al. Endothelin-3 regulates endothelin-1 production in cultured human endothelial cells. Hypertension 1991; 18: 304-315.

161. Nambi P, Pullen M, Kumar C. Identification of a novel endothelin receptor in *Xenopus Laevis* liver. Neuropeptides 1994; 26: 181-185.

162. Karne S, Jayawickreme CK, Lerner MR. Cloning and characterization of an endothelin-3 specific receptor (ET$_C$ receptor) from *Xenopus laevis* dermal melanophores. J Biol Chem 1993; 268: 19126-19133.

163. Emori T, Hirata Y, Marumo F. Specific receptors for endothelin-3 in cultured bovine endothelial cells and its cellular mechanism of action. FEBS Lett 1990; 263: 261-264.

164. Warner TD, Schmidt HH, Murad F. Interactions of endothelins and EDRF in bovine native endothelial cells: selective effects of endothelin 3. Am J Physiol 1992; H1600-H1605.

165. Takayanagi R, Kitazumi K, Takasaki C et al. Presence of nonselective type of endothelin receptor on vascular endothelium and its linkage to vasodilatation. FEBS Lett 1991; 282: 103-106.

166. Woodcock EA, Land S. Endothelin receptors in rat renal papilla with a high affinity for endothelin-3. Eur J Pharmacol-Mol Pharmacol Section 1991; 208: 255-260.

167. Kurihara T, Akimoto M, Kurokawa K et al. ET-3 sensitive reduction of tissue blood flow in rat liver. Life Sci 1992; 51: PL101-106.

168. Samson WK, Skala KD, Alexander BD et al. Pituitary site of action of endothelin: selective inhibition of prolactin release in vitro. Biochem Biophys Res Commun 1990; 169: 737-743.

169. Samson WK. The endothelin-A receptor subtype transduces the effects of the endothelins in the anterior pituitary gland. Biochem Biophys Res Commun 1992; 187: 590-595.

170. Hosada K, Nakao K, Arai H et al. Cloning and expression of human endothelin-1 receptor cDNA. FEBS Lett 1991; 287: 223-26.

171. Lin HY, Kaji EH, Winkel GK et al. Cloning and functional expression of a vascular smooth muscle endothelin 1 receptor. Proc Natl Acad Sci USA 1991; 88: 3185-3189.

172. Elshourbagy NA, Korman DR, Wu HL et al. Molecular characterization and regulation of the human endothelin receptors. J Biol Chem 1993; 268: 3873-3879.

173. Ogawa Y, Nakao K, Arai H et al. Molecular cloning of a non-isopeptide-selective human endothelin receptor. Biochem Biophys Res Commun 1991; 178: 248-255.

174. Mauzy C, Wu L-H, Egloff AM et al. Substitution of lysine-181 to aspartic acid in the third transmembrane region of the endothelin (ET) type B receptor selectively reduces its high affinity binding with ET-3 peptide. J Cardiovasc Pharmacol 1992; 20, Suppl 12: S5-S7.

175. Lee JA, Elliot JD, Sutiphong JA et al. Tyr-129 is important to the peptide ligand affinity and selectivity of human endothelin type A receptor. Proc Natl Acad Sci USA 1994; 91: 7164-7168.

176. Zhu G, Wu LH, Mauzy C et al. Replacement of lysine-181 by aspartic acid in the third transmembrane region of endothelin type B receptor reduces its affinity to endothelin peptides and sarafotoxin 6c without affecting G protein coupling. J Cell Biochem 1992; 50: 159-164.

SIGNAL TRANSDUCTION MECHANISMS OF THE ENDOTHELINS

Gillian A. Gray

In early reports, smooth muscle contraction evoked by the endothelium-derived peptidergic constrictor factor was said to be dependent on the presence of calcium (Ca^{2+}) in the extracellular medium.[1,2] This, together with the observation that contraction was prevented by blockers of dihydropyridine sensitive Ca^{2+} channels, led Yanagisawa et al to speculate that the new peptide endothelin might be an endogenous agonist of these channels.[3] However, subsequent studies have shown that ET-1 does not compete for binding of selective ligands to L-type Ca^{2+} channels and that the binding of ET-1 to its specific receptors is not inhibited by these ligands.[4-6] It is now recognized that the ETs, like many other peptides, bind to specific G-protein coupled receptors and activate a variety of cellular effector mechanisms (Fig. 5.1). The ETs stimulate short term changes in cell function, such as contraction of smooth muscle, but also longer term changes, such as stimulation of cell proliferation. The purpose of this chapter is to briefly review the signal transduction mechanisms through which the ETs bring about these changes. For fuller discussion of this topic the reader is referred to several reviews that deal exclusively with the cell signaling pathways of ET-1.[7,8]

G-PROTEINS

Guanine-nucleotide bindinG-proteins (G-proteins) are key components in cellular signaling processes in that they allow communication

Molecular Biology and Pharmacology of the Endothelins edited by Gillian A. Gray and David J. Webb. © 1995 R.G. Landes Company.

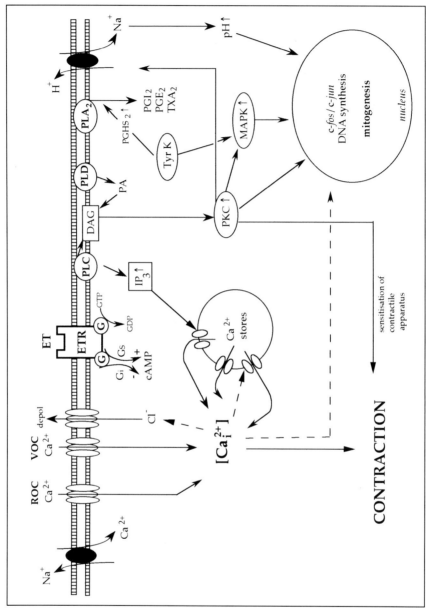

Fig. 5.1 Cellular signaling pathways of the endothelins. This diagram shows some of the pathways activated by ET receptors (ETR) that lead to contraction of vascular smooth muscle and to the mitogenic effect of the ETs. Evidence for linkage of these various pathways to G-proteins is reviewed in Table 5.1 and interactions between them are discussed in the text. ROC, receptor-operated calcium channel; VOC, voltage-operated calcium channel; G, G proteins; cAMP, cyclic AMP; PLC, phospholipase C; IP_3, inositol triphosphate; DAG, diacylglycerol; PLD, phospholipase D; PA, phosphatidic acid; PLA_2, phospholipase A_2; $PGHS_2$, type 2 prostaglandin endoperoxide synthase; Tyr K, tyrosine kinase; MAPK, mitogen-activated protein kinase; PKC, protein kinase C.

between receptors and effector mechanisms.[9,10] As discussed in chapter 3, cloning of the ET_A and ET_B receptors revealed that they belong to the superfamily of G-protein coupled 7-transmembrane domain receptors.[11,12] However, the pivotal role of G-proteins in regulating the effects of ET-1 had already been recognized prior to cloning. In several cell types, including cultured vascular smooth muscle cells,[13] mesangial cells[7] and ventricular myocytes,[14] ADP-ribosylation of G_i-proteins by pertussis toxin attenuated either the generation of cellular effectors or the actions of ET-1. Furthermore, the stable GTP analogue, GTP γs was found to potentiate the generation of effectors by ET-1.[7]

The G-proteins comprise a widely distributed and diverse family,[9,10] and it was soon evident that G-proteins other than the G_i subtype are involved in coupling of the ET receptors to intracellular signaling mediators. The G-proteins implicated in the generation of various cell effectors are discussed along with these effectors below. Some examples are summarized in Table 5.1. In fact it has now become clear that in many cells the ET receptors are capable of activating several signaling mechanisms via coupling to more than one G-protein. For example, in hepatocytes, changes in intracellular Ca^{2+} concentration that occur in response to ET-1 depend on consecutive activation of both G_q- and G_s-proteins that are coupled to ET_B receptors.[15] Also, in rabbit rectosigmoidal smooth muscle cells, different isotypes of the G_i-protein are implicated in the early transient and the late sustained phases of the contraction induced by ET-1.[16]

CALCIUM

In vascular smooth muscle, ET-1, like many other vasoconstrictor agents, elicits a biphasic increase in intracellular Ca^{2+}.[17] The first phase develops rapidly but is transient and depends on the mobilization of Ca^{2+} from intracellular stores. The second phase, in contrast, is slower to develop, but sustained and is dependent on the influx of Ca^{2+} from the extracellular space.

The inositol triphosphate (IP_3) sensitive Ca^{2+} store of the sarcoplasmic reticulum is a major source of Ca^{2+} mobilized by vasoconstrictor agents in smooth muscle.[18] As discussed below, IP_3 generation is an early event that follows activation of phospholipase C by ET-1. Pretreatment of cells with caffeine, which depletes the IP_3 sensitive store in the absence of extracellular Ca^{2+}, prevents the initial phase of Ca^{2+} release by ET-1 or SRTX S6b.[19,20] Ryanodine, which selectively inhibits Ca^{2+} release from the sarcoplasmic reticulum, also attenuates the initial ET-1 induced Ca^{2+} transient release.[21,22] IP_3 can release Ca^{2+} from both ryanodine-sensitive and ryanodine-insensitive stores, and both of these stores are mobilized by ET-1, in contrast to other constrictors like phenylephrine, which mobilizes Ca^{2+} exclusively from the ryanodine-sensitive store.[21] In cardiac myocytes, there is also evidence that ET-1 might release Ca^{2+} from IP_3 sensitive stores that lie outwith

Table 5.1. G-protein linked ET receptor signalling

G-Protein Implicated		Signal	Receptor Activated	Tissue Subtype	Ref
PTX-sensitive	G_i/G_o	cAMP ↓	ET_A	guinea-pig ventricular myocytes	35
			ET_B	transfected CHO cells	64
			ET_B	LLC-PK1 kidney epithelial cells	117
			ET_B	cultured bovine EC	65
			ET-3 selective ?	cultured bovine EC	66
		PLC/PI turnover	ET_A	astrocytes/glioma cells	130
		PLA_2/AA release	undefined	bovine iris sphincter smooth muscle	131
		Ca^{2+} influx (non-VOC)	ET_B	porcine EC in situ	132
		Ca^{2+} influx (VOC)	undefined	porcine coronary artery	133
		K^+ current	ET_A	rabbit atrial myocytes	134
PTX sensitive & insensitive components		PLC/PI turnover	ET_A	glioma cells	138
			ET_A	rat mesangial cells	7
PTX insensitive	G_s	cAMP ↑	ET_A	transfected CHO	64
			ET_A	cultured VSMC	65
		PLC/PI turnover	ET_A	human bronchial smooth muscle	135
		"	ET_A	cultured cerebellar granule cells	130
	$G_{q\alpha}$	"	ET_B	liver plasma membrane	15
	G_q	"	ET_B	rat cerebellum	136
		"	undefined	porcine coronary	133
	G_s	inhibition Ca^{2+} extrusion	ET_B	liver plasma membrane	15
		voltage-sensitive Ca^{2+} current	undefined	rabbit cardiomyocytes	137
		MAP kinase phosphorylation	undefined	cultured astrocytes	125

Signals activated by stimulation of ET receptors are classified according to their sensitivity to inhibition following ADP-ribosylation by pertussis toxin (PTX). Specific G-protein types involved are given where they have been identified. cAMP, adenylate cyclase; PLC, phospholipase C; PI, phosphatidyl inositol; PLA_2, phospholipase A_2; AA, arachidonic acid; VOC, voltage-operated Ca^{2+} channel; MAP kinase, mitogen-activated protein kinase; CHO, chinese hamster ovary; EC, endothelial cells; VSMC, vascular smooth muscle cells.

the sarcoplasmic reticulum.[23] Not all of the Ca^{2+} mobilized by ET-1 can be accounted for by IP_3 releasable stores. In cultured rat aortic smooth muscle cells, both ET-1 and ET-3 mobilize intracellular Ca^{2+} but only ET-1 stimulates an increase in IP_3.[24] An alternative mechanism of intracellular mobilization is via Ca^{2+} that has entered from outside of the cell (Ca^{2+}-induced Ca^{2+} release).[18] Observations that in some cells, both phases of the intracellular Ca^{2+} response to ET-1 can be blocked by antagonists of plasma membrane Ca^{2+} channels, suggest that this mechanism may also play a role in Ca^{2+} mobilization by ET-1.[25,26] In hepatocytes, in addition to mobilization of intracellular stores, ET-1 augments the cytosolic Ca^{2+} concentration through inhibition of Ca^{2+} extrusion by the plasma membrane pump.[15]

Calcium can enter cells from the extracellular space by several different mechanisms, including voltage-operated Ca^{2+} channels (VOCs), receptor-operated Ca^{2+} channels (ROCs) and by Na^+/Ca^{2+} exchange (reviewed in ref. 27). Blockade of the effects of ET-1 by antagonists of L-type VOCs shows that this is a major route for ET-1 evoked Ca^{2+} entry into various cells. Activation of VOCs by ET-1 in some, but not all, tissues takes place secondary to membrane depolarization.[4,28-30] The ionic basis for depolarization seems to be related to direct or indirect activation of Cl^- channels,[31-35] although other mechanisms such as blockade of ATP-sensitive K^+ channels[36] and increased intracellular Na^+ concentration by inhibition of Na^+/K^+-ATPase,[37,38] or by stimulation of the Na^+/H^+ antiporter[39] have also been implicated. In rat renal resistance arteries, membrane depolarization begins following the activation of a Ca^{2+}dependent Cl^- current and is terminated by activation of Ca^{2+} dependent K^+ channels.[40] Opening of Ca^{2+} dependent K^+ channels by ET-1 has been shown to occur in a number of other cells,[41,42] including smooth muscle, where it is thought that the resultant hyperpolarization might contribute to relaxation induced by ET-1.[42-45]

In several tissues, responses to ET-1 are attenuated by removal of Ca^{2+} from the extracellular medium or by addition of nonspecific cation channel blockers like Ni^{2+} or La^{3+}, but only marginally by Ca^{2+} channel antagonists.[30,46-48] In these circumstances, it is suggested that ET-1 might evoke Ca^{2+} influx through voltage-insensitive ROCs or via nonselective cation channels. Another possible route of entry for Ca^{2+} is by Na^+/Ca^{2+} exchange, and there is limited evidence that this mechanism might contribute to ET-1-induced Ca^{2+} uptake.[49,50]

Sensitivity of responses induced by ET-1 to removal of extracellular Ca^{2+} or to L-type Ca^{2+} channel antagonists varies between tissues but also within the same tissue, depending on the concentration of ET-1 that is applied.[30,48,51,52] One possible explanation for this heterogeneity is that, within one tissue, ET receptors with differing affinities for ET-1 are mediating the response to ET via separate signaling pathways. For example, high affinity receptors might mediate the opening of non-VOCs by low concentrations of ET-1, while a lower affinity receptor modulates opening of VOCs in response to higher concentrations of ET-1.[52] Varying contributions of intracellular Ca^{2+} to ET-1 evoked responses might also contribute to heterogeneity in sensitivity to blockade of Ca^{2+} entry from the extracellular space. In rat tracheal rings, contraction induced by the ET_B selective agonist, SRTX S6c, is attenuated to a far greater extent by removal of extracellular Ca^{2+}, than that induced by ET-1, which activates both ET_A and ET_B type receptors.[53] This might be explained by linkage of ET_A, but not ET_B receptors, to mobilization of intracellular Ca^{2+} in this tissue. In contrast, there is clear evidence that intracellular Ca^{2+} stores are released following stimulation of endothelial ET_B receptors.[54,55]

PHOSPHOLIPASE C

In vascular smooth muscle, stimulation of phospholipase C (PLC) activity leading to phosphatidyl inositol (PI) hydrolysis, IP_3 generation and mobilization of intracellular Ca^{2+},[56-58] and diacylglycerol (DAG)-mediated activation of protein kinase C (PKC),[39,59] is well established. ET-1 also activates PLC in other cell types, including nonvascular smooth muscle,[53,60] hepatocytes[15] and renal mesangial cells.[7] A number of studies have suggested that ET receptors are coupled to PLC activation via a pertussis toxin sensitive G_i-protein.[7,13,61] However, this is clearly not the case in all cell types, for example ET-1-induced PI hydrolysis is unaffected by pertussis toxin in rat fibroblasts,[62] rat aortic smooth muscle cells[63] and rat atrial tissue.[63] In these cells, ET-1 may be linked to PI hydrolysis via a G_q-protein, as is the case in hepatocytes.[15] Both ET_A and ET_B receptors couple to PI hydrolysis in transfected CHO cells,[64] as they do in many tissues, for example, ET_A receptors in smooth muscle cells[56] and cardiac myocytes,[61] and ET_B receptors in endothelial cells[65,66] and hepatocytes.[15] However ET receptors are not always coupled to PLC. For example, Little et al[24] showed that ET-1, but not ET-3, stimulates accumulation of IP_3 in cultured vascular smooth muscle cells. Similarly, in rat trachea, both ET-1 and SRTX S6c induce contraction, via ET_A and ET_B receptors consecutively, but only ET-1 increases IP_3 accumulation within the same concentration range at which it induces contraction.[53] These results suggest that ET_B receptors in vascular smooth muscle are not linked to PLC, a contention which is supported by the apparent absence of a role for PKC in ET_B receptor-mediated contraction (see below).[67]

PROTEIN KINASE C

ET_A receptor-mediated contraction of vascular smooth muscle by ET-1 is characteristically slow to develop and long-lasting.[1,3] Although unique among endogenous vasoconstrictor agents, this pattern is temporally similar to the contraction evoked by phorbol esters, exogenous activators of PKC.[68] It was, therefore, proposed that activation of PKC is involved in the sustained component of the ET-1 induced contraction.[59] A role for PKC as a cellular effector for ET-1 is supported by observations that ET-1 stimulates accumulation of DAG, the endogenous activator of PKC, in vascular smooth muscle cells[59] and facilitates translocation of several PKC isotypes from the cytosol to the membrane in these cells, as well as in cardiac myocytes and V6 glioma cells.[69-71] The PKC inhibitors staurosporine,[39,72] H-7,[73,74] calphostin C[75] and Ro 31-8220[67] attenuate ET-1-induced contraction of smooth muscle, as well as other actions of ET-1, like inhibition of platelet aggregation[76] and stimulation of atrial natriuretic peptide secretion from atrial cardiomyocytes.[77,78] With regard to the role of PKC in mediating sustained contraction of smooth muscle by ET-1, it is of interest to note that the rapidly developing and short-lasting contraction of smooth muscle mediated by ET_B receptors is not inhibited by the PKC inhibitor, Ro-31-8220.[67]

In vascular smooth muscle ET-1 evokes small, but sustained contractions in the absence of measurable increases in intracellular Ca^{2+} concentration.[21,30,72] It has been proposed that ET-1 potentiates the sensitivity of contractile elements to Ca^{2+} or by independent activation of the contractile elements.[21,79] An involvement of PKC in this process is suggested by the ability of PKC inhibitors to inhibit the Ca^{2+}-independent component of contraction in the porcine coronary artery.[80] PKC inhibitors also block enhancement of smooth muscle sensitivity to noradrenaline and angiotensin II by ET-1.[81] ET-1 phosphorylates a number of proteins in vascular smooth muscle that are PKC substrates, including vimentin and myosin light chain.[82] Furthermore, Nishimura et al[83] have shown that ET-1 increases the Ca^{2+} sensitivity of the myofilaments in rat mesenteric arteries through both phosphorylation of thin filament regulatory proteins and down-regulation of myosin light chain phosphatase.

In rat ventricular myocytes, PKC might have an additional mechanism that involves enhancement of Ca^{2+} entry through sarcolemmal T-type Ca^{2+} channels by ET-1.[84] Intracellular alkalization via stimulation of the Na^+/H^+ exchanger by ET-1 occurs via PKC in some[8,85] but not all cell types.[86,87] PKC is also implicated as a mediator of the long-term changes in cell function induced by ET-1, as discussed below.

PHOSPHOLIPASE A$_2$

Generation of prostacyclin (PGI_2) and thromboxane A2 (TxA_2) from perfused rat and guinea pig lungs provided the first evidence that ET-1 might stimulate the metabolism of arachidonic acid by phospholipase A_2 (PLA_2).[88] The actions of ET-1 are inhibited in several cell types by quinacrine, an inhibitor of PLA_2, but interpretation of these data is limited by the lack of selectivity of this compound.[89-91] However, inhibitors of the cyclo-oxygenase and lipoxygenase enzymes that prevent generation of PGI_2, TxA_2 and of the leukotrienes also modify responses to ET-1.[91,92] ET-1 stimulates generation of PGI_2 from endothelial cells via the ET_B receptor subtype, but can also activate PLA_2 via ET_A receptors in a number of cells.[7,13,56,93] The mechanism of PLA_2 activation by ET-1 has alternatively been proposed to be direct via a G-protein, as with alpha-adrenoreceptors[94] or indirect, by increasing the intracellular concentration of Ca^{2+}.[95,96] In CHO cells transfected with human ET receptors, ET-1 can activate PLA_2 via both ET_A and ET_B receptor subtypes.[93]

PHOSPHOLIPASE D

Phospholipase D catalyzes the hydrolysis of phospholipids to produce phosphatidic acid (PA), which can enhance Ca^{2+} influx, stimulate PLC activity and promote cell growth.[97] In addition, PKC can be activated by PA, following its conversion to DAG by a PA phosphohydrolase.[97] ET-1 increases PLD activity in a number of cells, including vascular smooth muscle,[98] osteoblasts[99] and in the SK-N-MC neuroblastoma cell line.[100] PLD is often associated with long-term signaling and it has therefore been suggested that it is a mediator of the mitogenic effect of ET-1,[101] with tyrosine kinase acting as an intermediary for PLD activation.[102] However, it may also modulate the short term signaling pathways of ET-1. For example, in SK-N-MC cells, PLD activation regulates stimulation of PLC by ET-1.[100]

TYROSINE KINASES

Protein tyrosine kinases (PTK), enzymes that phosphorylate proteins on tyrosine residues, can participate in the activation of PLD, and like PLD they are frequently associated with long-term changes in cell signalling.[101-103] The role of tyrosine phosphorylation of cellular proteins in the mitogenic effects of ET-1 is well documented.[104-106] Recent studies using novel inhibitors of PTK enzymes show that they can also influence short term signaling, such as that involved in contraction of smooth muscle.[107] One of these inhibitors, genistein, attenuates ET_B receptor-mediated contraction of rat tracheal smooth muscle by SRTX S6c (Fig. 5.2). Possible mechanisms of PTK include modulation of nonselective ion channels, as has been shown in coronary artery smooth muscle cells[108] or amplification of the IP_3 system, like that demonstrated in astrocytes overexpressing PTK.[109]

CYCLIC AMP

Depending on the tissue studied and the conditions used, levels of cyclic AMP (cAMP) can be increased or decreased by ET-1. For example, in endothelial cells, ET_B receptors mediate inhibition of forskalin-stimulated cAMP formation by ET-3.[65] Similarly in the rat tail artery and in guinea-pig ventricular myocytes, ET-1 decreases tissue levels of cAMP via activation of ET_A receptors.[35,110] In contrast, activation of ET_A receptors in cultured vascular smooth muscle cells by ET-1 leads to dose-dependent stimulation of cAMP formation.[65] These differences

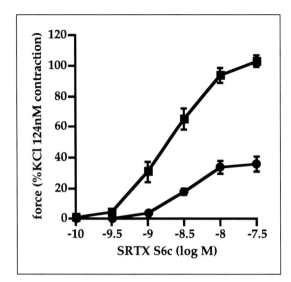

Fig. 5.2 Inhibition of ET_B receptor-mediated contraction of rat tracheal rings by the protein tyrosine kinase inhibitor, genistein. Contraction is induced by the ET_B receptor selective agonist, SRTX S6c in the absence (-■-) and the presence of genistein, 100 μM (-●-).

reflect linkage of ET receptors to adenylate cyclase via either a stimulatory G_s-protein or an inhibitory G_i-protein. Thus, the action of ET-1 on cellular cAMP level is governed not by the receptor subtype but rather by the type of G-protein to which the receptor is coupled and by the cell type in which it is expressed.

NITRIC OXIDE AND CYCLIC GMP

In addition to vasoconstriction via receptors on vascular smooth muscle, ET-1 can also induce vasodilatation by releasing the relaxing factors nitric oxide (NO) and PGI_2 from the vascular endothelium.[88] In isolated blood vessels and in cultured cells, ET-1, ET-3 and the ET_B receptor selective agonist SRTX S6c are equipotent in releasing relaxing factors from the endothelium, consistent with this effect being mediated via an ET_B type of receptor.[55,111-114] Inhibition of ET-3 induced liberation of intracellular Ca^{2+} and relaxation of rat aorta by the ET_B receptor antagonist, IRL 1038 confirms the role of ET_B receptors.[115] Endothelial ET_B receptors are coupled to PLC by a pertussis toxin sensitive G_i-protein[55] and liberation of Ca^{2+} from IP_3 sensitive intracellular stores is believed to lead to activation of the endothelial constitutive Ca^{2+}/calmodulin dependent NO synthase enzyme.[54,55] Recent evidence also implicates PTK-dependent pathways in liberation of NO via endothelial ET_B receptors.[54] On reaching the underlying smooth muscle cells, NO activates soluble guanylate cyclase leading to the formation of cyclic GMP and relaxation via a number of mechanisms that reduce cytosolic Ca^{2+} concentration.[116] ET-1 also stimulates formation of cGMP in kidney epithelial LLC-PK1 cells via ET receptors that are blocked by the ET_B receptor selective antagonist, BQ-788.[117] Unlike the endothelial cells, the receptors in these cells appear to be coupled to guanylate cyclase via a pertussis toxin insensitive G-protein.

NUCLEAR SIGNALING MECHANISMS

ET-1 stimulates DNA synthesis and proliferation in a number of cell types, including vascular smooth muscle cells,[102,118] glomerular mesangial cells,[103] thyrocytes,[119] melanocytes,[120] glial cells[121] and cardiac myocytes.[122] In most of these cells, ET-1 itself is a rather weak mitogen and addition of other growth factors is required in order to stimulate markers of proliferation, such as thymidine incorporation or expression of the proto-oncogenes, c-*fos* and c-*jun*. The growth factors with which ET-1 exhibits synergism include insulin, epidermal growth factor, fibroblast growth factor and platelet-derived growth factor.[7]

The role of protein kinases as mediators of the short term actions of ET-1 is discussed above. The signaling pathways that couple ET receptors to stimulation of DNA synthesis and proliferation are not completely elucidated, but there is evidence that several of these kinases, includinG-protein kinase C (PKC) and protein tyrosine kinase (PTK), also participate in this long term signaling process. Activation of mitogen-activated protein kinases (MAPK), a novel family of serine/threonine kinases, is an important step in the pathway linking the receptors of many growth factors to induction of cell growth.[123] ET-1 activates MAPK in several cell types.[103,124-126] In glomerular mesangial cells, depletion of PKC attenuates both activation of MAPK (p42-44^{mapk}) and induction of the proto-oncogene c-*fos* by ET-1.[126] Activation of p42 MAPK by ET-1 is also dependent on PKC in ventricular cardiomyocytes.[127] For full activation, MAPK require phosphorylation of tyrosine residues. In accordance with this PTK has been shown to contribute to mitogenic signaling and activation of MAPKs by ET-1 in mesangial cells[103] as does in astrocytes.[125] In vascular smooth muscle cells, phosphatidic acid formed as a result of PLD activation is implicated in the mitogenic effect of ET-1.[102] Both PKC and PTK are required for activation of PLD by ET-1 in these cells.[102] As discussed above, ET-1 elicits an increase in cytosolic Ca^{2+} concentration through a number of mechanisms. In primary cultures of rat cerebellum, ET-1 causes a Ca^{2+} dependent increase in [^{3}H] thymidine incorporation, suggesting that Ca^{2+} is a requirement for cellular proliferation.[121] The role of Ca^{2+} is most likely complex since in vascular smooth muscle and mesangial cells, induction of the proto-oncogenes c-*fos* and c-*jun* by ET-1 is enhanced rather than inhibited by blockade of Ca^{2+} entry using the Ca^{2+} channel antagonist, manidipine.[26]

The mitogenic effect of ET-1 has frequently been linked to activation of the ET_A receptor subtype.[103,128] However recent studies, using CHO cells transfected with ET_A or ET_B receptors, show that the MAPK cascade and cellular proliferation can be stimulated by both ET_A and ET_B receptor subtypes.[129]

REFERENCES

1. Hickey KA, Rubanyi GM, Paul RJ et al. Characterization of a coronary vasoconstrictor produced by cultured endothelial cells. Am J Physiol 1985; 248: C550-C556.
2. Gillespie MN, Owasoyo JO, McMurty IF et al. Sustained coronary vasoconstriction provoked by a peptidergic substance released from endothelial cells in culture. J Pharmacol Exp Ther 1986; 236: 339-343.
3. Yanagisawa M, Kurihara H, Kimura S et al. A novel potent vasoconstrictor peptide produced by vascular endothelial cells. Nature 1988; 332: 411-415.
4. Kasuya Y, Ishikawa T, Yanagisawa M et al. Mechanism of contraction to endothelin in isolated porcine coronary artery. Am J Physiol 1989; 257: H1828-H1835.
5. Hirata Y, Yoshimi H, Takaichi S et al. Binding and receptor down-regulation of a novel vasoconstrictor endothelin in cultured rat vascular smooth muscle cells. FEBS Lett 1988; 239: 13-17.
6. Clozel M, Fischli W, Guilly C. Specific binding of endothelin on human vascular smooth muscle cells in culture. J Clin Invest 1989; 83: 1758-1761.
7. Simonson MS, Dunn MJ. Cellular signaling by peptides of the endothelin gene family. FASEB J 1993; 4: 2829-3000.
8. Brock TA, Danthaluri NR. Cellular actions of endothelin in vascular smooth muscle. In: Rubanyi GM, ed(s). Endothelin. New York: Oxford University Press, 1992: 103-124.
9. Freissmuth M, Casey PJ, Gilman AG. G-proteins control diverse pathways of transmembrane signaling. FASEB J 1989; 3: 2125-2131.
10. Milligan G. Tissue distribution and subcellular location of guanine nucleotide regulatory proteins: implications for cellular signalling. Cell Signal 1989; 1: 411-419.
11. Arai H, Hori S, Aramori I et al. Cloning and expression of a cDNA encoding an endothelin receptor. Nature 1990; 348: 730-732.
12. Sakurai T, Yanagisawa M, Takuwa Y et al. Cloning of a cDNA encoding a non-isopeptide selective subtype of the endothelin receptor. Nature 1990; 348: 732-735.
13. Reynolds EE, Mok LL, Kurokawa S. Phorbol ester dissociates endothelin stimulated phosphoinositide hydrolysis and arachidonic acid release in vascular smooth muscle cells. Biochem Biophys Res Commun 1989; 160: 868-873.
14. Kelly RA, Eid H, Kramer B et al. Endothelin enhances the contractile responsiveness of adult rat ventricular myocytes to calcium via a pertussis toxin sensitive pathway. J Clin Invest 1990; 86: 1164-1171.
15. Jouneaux C, Mallat A, Serradeil Le Gal C et al. Coupling of endothelin B receptors to the calcium pump and phospholipase C via Gs and Gq in rat liver. J Biol Chem 1994; 269: 1845-1851.
16. Bitar KN, Stein S, Omann GM. Specific G-proteins mediate endothelin-induced contraction. Life Sci 1992; 50: 2119-2124.

17. Sakata K, Ozaki H, Kwon S et al. Effects of endothelin on the mechanical activity and cytosolic calcium levels of various types of smooth muscle. Br J Pharmacol 1989; 98: 483-492.

18. Van Breemen C, Cauvin C, Johns A et al. Ca^{2+} regulation of vascular smooth muscle. Federation Proc 1986; 45: 2746-2751.

19. Kai H, Kanaide H, Nakamura M. Endothelin-sensitive intracellular Ca^{2+} store overlaps with caffeine-sensitive one in rat aortic smooth muscle cells in primary culture. Biochem Biophys Res Commun 1989; 158: 235-243.

20. Watanabe C, Hirano K, Kanaide H. Role of extracellular and intracellular sources of Ca^{2+} in sarafotoxin S6b-induced contraction of strips of the rat aorta. Br J Pharmacol 1993; 108: 30-37.

21. Huang X-N, Hisayama T, Takayanagi I. Endothelin-1 induced contraction of rat aorta: contributions made by Ca^{2+} influx and activation of contractile apparatus associated with no change in cytoplasmic Ca^{2+} level. Naunyn-Schmeideberg's Arch Pharmacol 1990; 341: 80-87.

22. Wagner Mann C, Sturek M. Endothelin mediates calcium influx and release in porcine coronary smooth muscle cells. Am J Physiol 1991; 260: C771-C777.

23. Vigne P, Breittmayer JP, Frelin C. Thapsigargin, a new inotropic agent, antagonizes action of endothelin-1 in rat atrial cells. Am J Physiol 1992; 263: H1689-H1694.

24. Little PJ, Neylon CB, Tkachuk VA et al. Endothelin-1 and endothelin-3 stimulate calcium mobilization by different mechanisms in vascular smooth muscle. Biochem Biophys Res Commun 1992; 183: 694-700.

25. Gardner JP, Tokudome G, Tomonari H et al. Endothelin-induced calcium responses in vascular smooth muscle cells. Am J Physiol 1992; 262: C148-C155.

26. Huang S, Simonson MS, Dunn MJ. Manidipine inhibits endothelin-1 induced Ca^{2+} signalling but potentiates endothelins effect on c-fos and c-jun induction in vascular smooth muscle cells. Am Heart J 1993; 125: 589-597.

27. Spedding M, Paoletti R. Classification of calcium channels and the sites of action of drugs modifying channel function. Pharmacol Rev 1992; 44: 363-376.

28. Van Rhenterghem C, Vigne P, Barhanin J et al. Molecular mechanism of action of the vasoconstrictor peptide endothelin. Biochem Biophys Res Commun 1988; 157: 977-985.

29. Lee HK, Leikauf GD, Sperelakis N. Electromechanical effects of endothelin on ferret bronchial and tracheal smooth muscle. J Appl Physiol 1990; 68: 417-420.

30. Wallnofer A, Weir S, Ruegg U et al. The mechanism of endothelin-1 as compared with other agonists in vascular smooth muscle. J Cardiovasc Pharmacol 1989; 13, suppl 5: S23-S31.

31. Iijima K, Lin L, Nasjletti A et al. Intracellular signalling pathway of endothelin-1. J Cardiovasc Pharmacol 1991; 17, suppl 7: S119-S121.

32. Takenaka T, Epstein M, Forster H et al. Attenuation of endothelin ef-

fects by a chloride channel inhibitor, indanyloxyacetic acid. Am J Physiol 1992; 262: F799-F806.

33. Van Renterghem C, Lazdunski M. Endothelin and vasopressin activate low conductance chloride channels in aortic smooth muscle cells. Pflug Arch 1993; 425: 156-1163.

34. Klockner U, Isenberg G. Endothelin depolarizes myocytes from porcine coronary and human mesenteric arteries through a Ca-activated chloride current. Pflug Arch 1991; 418: 1-2.

35. James AF, Xie L-H, Fujitani Y et al. Inhibition of the cardiac protein kinase A-dependent chloride conductance by endothelin-1. Nature 1994; 370: 297-300.

36. Miyoshi Y, Nakaya Y, Wakatsuki T et al. Endothelin blocks ATP-sensitive K^+ channels and depolarizes smooth muscle cells of porcine coronary artery. Circ Res 1992; 70: 612-616.

37. Meyer-Lehnart H, Wanning C, Predel HG et al. Effects of endothelin on sodium transport mechanisms:potential role in cellular calcium mobilization. Biochem Biophys Res Commun 1989; 163: 458-465.

38. Pecci A, Cozza EN, Devlin M et al. Endothelin-1 stimulation of aldosterone and zona glomerulosa-ouabain-sensitive sodium/potassium-ATPase. J Steroid Biochem Mol Biol 1994; 50: 49-53.

39. Danthaluri NR, Brock TA. Endothelin-receptor coupling mechanisms in vascular smooth muscle: a role for protein kinase C. J Pharmacol Exp Ther 1990; 254: 393-399.

40. Gordienko DV, Clausen C, Goligorsky MS. Ionic currents and endothelin signaling in smooth muscle cells from rat renal resistance arteries. Am J Physiol 1994; 266: F325-F341.

41. Supattapone S, Ashley CC. Endothelin increases rubidium uptake through calcium-activated potassium channels in C6 glioma cells. Ann New York Acad Sci 1991; 633: 597-598.

42. Hu S, Kim HS, Jeng AY. Dual action of endothelin-1 on the Ca^{2+}-activated K^+ channel in smooth muscle cells of porcine coronary artery. Eur J Pharmacol 1991; 194: 31-36.

43. Fulginiti JI, Cohen MM, Moreland RS. Endothelin differentially affects rat gastric longitudinal and circular smooth muscle. J Pharmacol Exp Ther 1993; 265: 1413-1420.

44. Lin WW, Lee CY. Intestinal relaxation by endothelin isopeptides: involvement of Ca^{2+}- activated K^+ channels. Eur J Pharmacol 1992; 219: 355-360.

45. Borg C, Mondot S, Mestre M et al. Nicorandil: differential contribution of K^+ channel opening and guanylate cyclase stimulation to its vasorelaxant effects on various endothelin-1-contracted arterial preparations: comparison to aprikalim (RP 52891) and nitroglycerin. J Pharmacol Exp Ther 1991; 259: 526-534.

46. Blackburn K, Highsmith RF. Nickel inhibits endothelin-induced contractions of vascular smooth muscle. Am J Physiol 1990; 258: C1025-C1030.

47. Stefan M, Russell JA. Signal transduction in endothelin-induced contraction of rabbit pulmonary artery. Pulmonary Pharmacology 1989; 3: 1-7.

48. Stasch JP, Kazda S. Endothelin 1 induced vascular contractions: interactions with drugs affecting the calcium channel. J Cardiovasc Pharmacol 1989; 13, Suppl 5: S63-S66.

49. Borges R, Carter DV, Von Grafenstein H et al. Ionic requirements of the endothelin response in aorta and portal vein. Circ Res 1989; 65: 265-271.

50. Criscione L, Nellis P, Riniker B et al. Reactivity and sensitivity of mesenteric vascular beds and aortic rings of spontaneously hypertensive rats to endothelin: effects of calcium entry blockers. Br J Pharmacol 1990; 100: 31-36.

51. Suzuki Y, Tanoi C, Shibuya M et al. Different utilization of Ca^{2+} in the contractile action of endothelin-1 on cerebral, coronary and mesenteric arteries of the dog. Eur J Pharmacol 1992; 219: 401-408.

52. Advenier C, Sarria B, Naline E et al. Contractile activity of three endothelins (ET-1, ET-2 and ET-3) on the human isolated bronchus. Br J Pharmacol 1990; 100: 168-172.

53. Henry PJ. Endothelin-1 (ET-1)-induced contraction in rat isolated trachea: involvement of ET_A and ET_B receptors and multiple signal transduction systems. Br J Pharmacol 1993; 110: 435-441.

54. Tsukuhara H, Ende H, Magazine HI et al. Molecular and functional characterization of the non-isopeptide-selective ET_B receptor in endothelial cells. J Biol Chem 1994; 269: 21778-21785.

55. Hirata Y, Emori T, Eguchi S et al. Endothelin receptor subtype B mediates synthesis of nitric oxide by cultured bovine endothelial cells. J Clin Invest 1993; 91: 1367-1373.

56. Resink TJ, Scott-Burden T, Bühler FR. Endothelin stimulates phospholipase C in cultured vascular smooth muscle cells. Biochem Biophys Res Commun 1988; 157: 1360-1368.

57. Miasaro N, Yamamoto H, Kanaide H et al. Does endothelin mobilize calcium from intracellular store sites in rat aortic vascular smooth muscle cells in primary culture? Biochem Biophys Res Commun 1988; 156: 312-317.

58. Marsden PA, Danthaluri NR, Brenner BM et al. Endothelin action on vascular smooth muscle involves inositol triphosphate and calcium mobilization. Biochem Biophys Res Commun 1989; 158: 86-93.

59. Griendling KK, Tsuda T, Alexander RW. Endothelin stimulates diacylglycerol accumulation and activates protein kinase C in cultured vascular smooth muscle cells. J Biol Chem 1989; 264: 8237-8240.

60. Garcia-Pascual A, Persson K, Holmquist F et al. Endothelin-1-induced phosphoinositide hydrolysis and contraction in isolated rabbit detrusor and urethral smooth muscle. Gen Pharmacol 1993; 24: 131-138.

61. Hilal-Dandan R, Urasawa K, Brunton LL. Endothelin inhibits adenylate cyclase and stimulates phosphoinositide hydrolysis in adult cardiac myocytes. J Biol Chem 1992; 267: 10620-4.

62. Muldoon L, Rodland KD, Forsythe ML et al. Stimulation of phosphatidylinositol hydrolysis, diacylglycerol release and gene expression in response to endothelin, a potent new agonist for fibroblasts and smooth muscle cells. J Biol Chem 1989; 264: 8529-8536.

63. Takuwa Y, Kasuya Y, Takuwa N et al. Endothelin receptor is coupled to phospholipase C via a pertussis toxin-insensitive guanine nucleotide-binding regulatory protein in vascular smooth muscle. J Clin Invest 1990; 85: 653-658.

64. Aramori I, Nakanishi S. Coupling of two ET receptor subtypes to differing signal transduction in transfected chinese hamster ovary cells. J Biol Chem 1992; 267: 12468-12474.

65. Eguchi S, Hirata Y, Imai T et al. Endothelin receptor subtypes are coupled to adenylate cyclase via different guanyl nucleotide-bindinG-proteins in vasculature. Endocrinology 1993; 132: 524-529.

66. Emori T, Hirata Y, Kanno K et al. Endothelin-3 stimulates production of endothelium-derived nitric oxide via phosphoinositide breakdown. Biochem Biophys Res Commun 1991; 174: 228-235.

67. Gray GA, Löffler BM, Clozel M. Characterization of endothelin receptors mediating contraction of the rabbit saphenous vein. Am J Physiol 1994; 266: H959-H966.

68. Rasmussen H, Takuwa Y, Park S. Protein kinase C in the regulation of smooth muscle contraction. FASEB J 1987; 1: 177-185.

69. Lee TS, Chao T, Hu KQ et al. Endothelin stimulates a sustained 1,2 diacylglycerol and protein kinase C activation in bovine aortic smooth muscle cells. Biochem Biophys Res Commun 1989; 162: 381-386.

70. Bogoyevitch MA, Parker PJ, Sugden PH. Characterization of protein kinase C isotype expression in adult rat heart: protein kinase C-epsilon is a major isotype present, and it is activated by phorbol esters, epinephrine, and endothelin. Circ Res 1993; 72: 757-767.

71. Chen CC. Protein kinase C alpha, delta, epsilon and zeta in C6 glioma cells. TPA induces translocation and down-regulation of conventional and new PKC isoforms but not atypical PKC zeta. FEBS Lett 1993; 332: 1-2.

72. Ohlstein EH, Horohonich S, Hay DWP. Cellular mechanisms of endothelin in rabbit aorta. J Pharmacol Exp Ther 1989; 250: 548-555.

73. Murray MA, Faraci FM, Heistad DD. Effect of protein kinase C inhibitors on endothelin- and vasopressin- induced constriction of the rat basilar artery. Am J Physiol 1992; 263: H1643-H1649.

74. Sugiura M, Inagami T, Hare GMT et al. Endothelin action: inhibition by a protein kinase C inhibitor and involvement of phosphoinositols. Biochem Biophys Res Commun 1989; 168: 170-176.

75. Shimamoto H, Shimamoto Y, Kwan CY et al. Participation of protein kinase C in endothelin-1-induced contraction in rat aorta: studies with a new tool, calphostin C. Br J Pharmacol 1992; 107: 282-287.

76. Pietraszek MH, Takada Y, Takada A. Endothelins inhibit serotonin-induced platelet aggregation via a mechanism involvinG-protein kinase C. Eur J Pharmacol 1992; 219: 289-293.

77. Irons CE, Murray SF, Glembotski CC. Identification of the receptor subtype responsible for endothelin-mediated protein kinase C activation and atrial natriuretic factor secretion from atrial myocytes. J Biol Chem 1993; 268: 23417-23421.

78. Pitkanen M, Mantymaa P, Ruskoaho H. Staurosporine, a protein kinase C inhibitor, inhibits atrial natriuretic peptide secretion induced by sarafotoxin, endothelin and phorbol ester. Eur J Pharmacol 1991; 195: 307-315.

79. Itoh H, Higuchi H, Hiraoka N et al. Contraction of rat thoracic aorta strips by endothelin-1 in the absence of extracellular Ca^{2+}. Br J Pharmacol 1991; 104: 847-852.

80. Kodama M, Kanaide H, Abe S et al. Endothelin-induced Ca^{2+} independent contraction of the porcine coronary artery. Biochem Biophys Res Commun 1989; 160: 1302-1308.

81. Henrion D, Laher I. Potentiation of norepinephrine-induced contractions by endothelin-1 in the rabbit aorta. Hypertension 1993; 22: 78-83.

82. Tsuda T, Griendling KK, Ollerenshaw JD et al. Angiotensin-II- and endothelin-induced protein phosphorylation in cultured vascular smooth muscle cells. J Vasc Res 1993; 30: 241-249.

83. Nishimura J, Moreland S, Hee YA et al. Endothelin increases myofilament Ca^{2+} sensitivity in alpha-toxin-permeabilized rabbit mesenteric artery. Circ Res 1992; 71: 951-959.

84. Furukawa T, Ito H, Nitta J et al. Endothelin-1 enhances calcium entry through T-type calcium channels in cultured neonatal rat ventricular myocytes. Circ Res 1992; 71: 1242-1253.

85. Kramer BK, Smith TW, Kelly RA. Endothelin and increased contractility in adult rat ventricular myocytes: role of intracellular alkalosis induced by activation of the protein kinase C-dependent Na^+-H^+ exchanger. Circ Res 1991; 68: 269-279.

86. Vigne P, Frelin C. Endothelins activate phospholipase A_2 in brain capillary endothelial cells. Brain Research 1994; 651: 342-344.

87. Winkel GK, Sardet C, Pouyssegur J et al. Role of cytoplasmic domain of the Na^+/H^+ exchanger in hormonal activation. J Biol Chem 1993; 268: 3396-3400.

88. De Nucci G, Thomas R, D'Orleans-Juste P et al. Pressor effects of circulating endothelin are limited by its removal in the pulmonary circulation and by the release of prostacyclin and endothelium-derived relaxing factor. Proc Natl Acad Sci USA 1988; 85: 9797-9800.

89. Ahmed A, Cameron IT, Ferriani RA et al. Activation of phospholipase A_2 and phospholipase C by endothelin-1 in human endometrium. J Endocrinol 1992; 135: 383-390.

90. Kozuka M, Ito T, Hirose S et al. Endothelin action on rat uterus is inhibited by an inhibitor of protein kinase C and by inhibitors of the phospholipase A_2-arachidonic acid -lipoxygenase pathway. Biomed Res 1990; 11: 287-289.

91. Resink TJ, Scott-Burden T, Buhler FR. Activation of phospholipase A_2 by endothelin in cultured vascular smooth muscle cells. Biochem Biophys Res Commun 1989; 158: 279-286.

92. Ninomaya H, Uchida Y, Saotome M et al. Endothelins constrict guinea-pig airways by multiple mechanisms. J Pharmacol Exp Ther 1992; 262: 570-576.

93. Schramek H, Wang Y, Konieczkowski M et al. Endothelin-1 stimulates cytosolic phospholipase A_2 in chinese hamster ovary cells stably expressing the human ET_A or ET_B receptor subtype. Biochem Biophys Res Commun 1994; 199: 992-997.

94. Burch RM, Luini A, Axelrod J. Phospholipase A_2 and phospholipase C are activated by distinct GTP-bindinG-proteins in response to alpha 1-adrenergic stimulation in FRTL5 thyroid cells. Proc Natl Acad Sci USA 1986; 83: 7201-7205.

95. Azelrod J, Burch RM, Jelsema CL. Receptor-mediated activation of phospholipase A_2 via GTP-bindinG-proteins arachidonic acid and its metabolites as second messengers. Trends Neurol Sci 1988; 11: 117-123.

96. Barnett RL, Ruffini L, Hart D et al. Mechanism of endothelin activation of phospholipase A_2 in rat renal medullary interstitial cells. Am J Physiol 1994; 266: F46-F56.

97. Shukla SD, Halkenda SP. Phospholipase D in cell signalling and its relationship to phospholipase C. Life Sci 1991; 48: 851-866.

98. Liu Y, Geisbuhler B, Jones AW. Activation of multiple mechanisms including phospholipase D by endothelin-1 in rat aorta. Am J Physiol 1992; 262: C941-C949.

99. Suzuki A, Oiso Y, Kozawa O. Effect of endothelin-1 on phospholipase D activity in osteoblast-like cells. Mol Cell Endocrinol 1994; 105: 193-196.

100. Challiss RAJ, Wilkes LC, Patel V et al. Phospholipase D activation regulates endothelin-1 stimulation of phosphoinositide-specific phospholipase C in SK-N-MC cells. FEBS Lett 1993; 327: 157-160.

101. Boarder MR. A role for phospholipase D in control of mitogenesis. Trends Pharmacol Sci 1994; 15: 57-62.

102. Wilkes LC, Patel V, Purkiss JR et al. Endothelin-1 stimulated phospholipase D in A10 vascular smooth muscle derived cells is dependent on tyrosine kinase: evidence for involvement in stimulation of mitogenesis. FEBS Lett 1993; 322: 147-150.

103. Simonson MS, Herman WH. Protein kinase C and tyrosine kinase activity contribute to mitogenic signaling by endothelin: cross talk between G-protein-coupled receptors and pp60c-src. J Biol Chem 1993; 268: 9347-9357.

104. Cazaubon SM, Ramos-Morales F, Fischer S et al. Endothelin induces tyrosine phosphorylation and GRB2 association of Shc in astrocytes. J Biol Chem 1994; 269: 24805-24809.

105. Schaller MD, Parsons JT. Focal adhesion kinase: an integrin-linked protein tyrosine kinase. Trends Cell Biol 1993; 3: 258-262.

106. Zachary I, Sinnett-Smith J, Rozengurt E. Bombesin, vasopressin, and endothelin stimulation of tyrosine phosphorylation in Swiss 3T3 cells. Identification of a novel tyrosine kinase as a major substrate. J Biol Chem 1992; 267: 19031-19034.

107. Di Salvo J, Steusloff A, Semenchuk L et al. Tyrosine kinase inhibitors suppress agonist-induced contraction in smooth muscle. Biochem Biophys Res Commun 1993; 190: 968-974.

108. Minami K, Fuzukawa K, Inuoe I. Regulation of a non-selective cation channel of cultured porcine coronary artery smooth muscle cells by tyrosine kinase. Pflügers Arch 1994; 426: 254-257.

109. Mattingly RR, Wasilenko WJ, Woodring PJ et al. Selective amplification of endothelin-stimulated inositol 1,4,5- trisphosphate and calcium signaling by v-src transformation of Rat-1 fibroblasts. J Biol Chem 1992; 267: 7470-7477.

110. Yang MCM, Tu MS, Chou CK et al. cAMP and vascular action of endothelin. Pharmacology 1991; 42: 252-256.

111. Edwards RM, Pullen M, Nambi P. Activation of endothelin ET_B receptors increases glomerular cGMP via an L-arginine-dependent pathway. Am J Physiol 1992; 263: F1020-F1025.

112. Karaki H, Sudjarwo SA, Hori M et al. Induction of endothelium-dependent relaxation in the rat aorta by IRL 1620, a novel and selective agonist at the endothelin ET_B receptor. Br J Pharmacol 1993; 109: 486-490.

113. Moritoki H, Miyano H, Takeuchi S et al. Endothelin-3-induced relaxation of rat thoracic aorta: a role for nitric oxide formation. Br J Pharmacol 1993; 108: 1125-1130.

114. Pinheiro JMB, Malik AB. Mechanisms of endothelin-1-induced pulmonary vasodilatation in neonatal pigs. J Physiol 1993; 469: 739-752.

115. Sudjarwo SA, Hori M, Karaki H. Effect of endothelin-3 on cytosolic calcium level in vascular endothelium and on smooth muscle contraction. Eur J Pharmacol 1992; 229: 137-142.

116. Lincoln TM. Cyclic GMP and mechanisms of vasodilatation. Pharmac Therap 1989; 41: 479-502.

117. Ozaki S, Ihara M, Saeki T et al. Endothelin ET_B receptors couple to two distinct signaling pathways in porcine kidney epithelial LLC-PK1 cells. J Pharmacol Exp Ther 1994; 270: 1035-1040.

118. Komuro I, Kurihara H, Sugiyama T et al. Endothelin stimulates c-*fos* and c-*myc* expression and proliferation of vascular smooth muscle cells. FEBS Lett 1988; 238: 249-252.

119. Eguchi K, Kawakami A, Nakashima M et al. Stimulation of mitogenesis in human thyroid epithelial cells by endothelin. Acta Endocrinol 1993; 128: 215-220.

120. Imokawa G, Yada Y, Miyagishi M. Endothelins secreted from human keratinocytes are intrinsic mitogens for human melanocytes. J Biol Chem 1992; 267: 24675-24680.

121. Supattapone S, Simpson A, Ashley C. Free calcium rise and mitogenesisin glial cells caused by endothelin. Biochem Biophys Res Commun 1989; 165: 1115-1122.

122. Suzuki T, Hoshi H, Mitsui Y. Endothelin stimulates hypertrophy and contractility of neonatal rat cardiac myocytes in a serum free medium. FEBS Lett 1990; 268: 149-151.

123. Pelech SL, Sanghera JS. Mitogen-activated protein kinases: versatile transducers for cell signalling. Trends Biochem Sci 1992; 17: 233-238.

124. Wang YZ, Simonson MS, Pouyssegur J et al. Endothelin rapidly stimu-

lates mitogen-activated protein kinase activity in rat mesangial cells. Biochem J 1992; 287: 589-594.

125. Cazaubon S, Parker PJ, Strosberg AD et al. Endothelins stimulate tyrosine phosphorylation and activity of p42 /mitogen-activated protein kinase in astrocytes. Biochem J 1993; 293: 381-386.

126. Simonson MS, Wang Y, Jones JM et al. Protein kinase C regulates activation of mitogen-activated protein kinase and induction of proto-oncogene c-fos by endothelin-1. J Cardiovasc Pharmacol 1992; 20, Suppl 12: S29-S32.

127. Bogoyevitch MA, Glennon PE, Sugden PH. Endothelin-1, phorbol esters and phenylephrine stimulate MAP kinase activities in ventricular cardiomyocytes. FEBS Lett 1993; 317: 271-275.

128. Kitagawa N, Tsutsumi K, Niwa M et al. A selective endothelin ET_A antagonist, BQ-123, inhibits ^{125}I-endothelin-1 (^{125}I-ET-1) binding to human meningiomas and antagonizes ET-1-induced proliferation of meningioma cells. Cell Molecul Neurobiol 1994; 14: 105-118.

129. Wang Y, Ros PM, Webb ML et al. Endothelins stimulate mitogen-activated protein kinase cascade through either ET_A or ET_B. Am J Physiol 1994; 267: C1130-C1135.

130. Chuang DM, Lin WW, Lee CY. Endothelin-induced activation of phosphoinositide turnover, calcium mobilization, and transmitter release in cultured neurons and neurally related cell types. J Cardiovasc Pharmacol 1991; 17, Suppl. 7: S85-S88.

131. Yousufzai SY, Abdel-Latif AA. Involvement of a pertussis toxin-sensitive G-protein-coupled phospholipase A_2 in agonist-stimulated arachidonic acod release in membranes isolated from bovine iris sphincter smooth muscle. Membr Biochem 1993; 10: 29-42.

132. Aoki H, Kobayashi S, Nishimura J et al. Sensitivity of G-protein involved in endothelin-1-induced Ca^{2+} influx to pertussis toxin in porcine endothelial cells in situ. Br J Pharmacol 1994; 111: 989-996.

133. Kasuya Y, Takuwa Y, Yanagisawa M et al. A pertussis toxin-sensitive mechanism of endothelin action in porcine coronary artery smooth muscle. Br J Pharmacol 1992; 107: 456-462.

134. Ono K, Tsujimoto G, Sakamoto A et al. Endothelin-A receptor mediates cardiac inhibition by regulating calcium and potassium currents. Nature 1994; 370: 301-304.

135. Mattoli S, Soloperto M, Mezzetti M et al. Mechanisms of calcium mobilization and phosphoinositide hydrolysis in human bronchial smooth muscle cells by endothelin 1. American Journal of Respiratory Cell, Molecular Biology 1991; 5: 424-30.

136. Sokolovsky M. Endothelin receptors in rat cerebellum: activation of phosphoinositide hydrolysis is transduced by multiple G-proteins. Cellular Signalling 1993; 5: 473-483.

137. Lauer MR, Gunn MD, Clusin WT. Endothelin activates voltage-dependent Ca^{2+} current by a G-protein-dependent mechanism in rabbit cardiac myocytes. J Physiol (Lond) 1992; 448: 729-747.

138. Gusovsky F. Endothelin-elicited stimulation of phospholipase C is mediated by guanine nucleotide bindinG-proteins. Eur J Pharmacol-Mol Pharmacol 1992; 225: 339-345.

================ CHAPTER 6 ================

CARDIOVASCULAR PHARMACOLOGY OF THE ENDOTHELINS

Paula J.W. Smith, William G. Haynes, David J. Webb

The widespread expression of mRNA for members of the endothelin (ET) family, and the distribution of their receptors, suggests that these peptides may play an important role in local regulation of the cardiovascular system. ET-1, the predominant isopeptide of the family, is a potent vasoconstrictor with a characteristically sustained action. ET-1 also has inotropic and mitogenic properties, influences salt and water homeostasis, stimulates generation of renin, angiotensin II, aldosterone and epinephrine and increases sympathetic activity. Together these actions promote vasoconstriction and increase blood pressure. ET-1 stimulates generation of endothelium-derived dilator substances that act to modulate its direct vasoconstrictor actions and probably account for the transient vasodilatation observed after bolus injections. ET-1 appears to be primarily a locally acting substance because circulating plasma concentrations are usually not sufficient to elicit direct vasoconstriction. However, concentrations at the interface between endothelial cells and vascular smooth muscle cells are probably higher, because more ET-1 is secreted abluminally than luminally.

This chapter reviews the cardiac, vascular and renal pharmacology of ET as demonstrated in vivo in animals and humans, concentrating predominantly on ET-1. In addition, it highlights those aspects of endocrine and central nervous system pharmacology that are relevant to the cardiorenal actions of ET. The potential physiological significance of ET in cardiovascular regulation is also addressed. Other aspects

Molecular Biology and Pharmacology of the Endothelins edited by Gillian A. Gray and David J. Webb. © 1995 R.G. Landes Company.

of the pharmacology of ET, including the respiratory, central nervous, endocrine, and gastrointestinal systems, are discussed only briefly because these specialized areas have been extensively reviewed elsewhere.[1-3]

BLOOD PRESSURE

PRESSOR RESPONSES

In their original paper, Yanagisawa et al[4] showed that intravenous administration of a bolus dose of ET-1 markedly increased blood pressure in chemically denervated rats (Fig. 6.1). This pressor effect was sustained for more than 60 minutes, in contrast to the brief effects of all other endogenous vasoconstrictor substances. ET-2 and ET-3 have also been shown to increase blood pressure in the rat,[5] with ET-3 evoking the least response and having the shortest action. Similar pressor responses to the ETs have been demonstrated subsequently by other investigators in the rat,[5-7] guinea pig,[8,9] rabbit,[10] dog,[11] pig,[12] and goat[13] as well as in nonmammalian species.[14] In the dog, infusion of lower doses of ET-1, sufficient to double plasma concentrations of the peptide, do not increase blood pressure, but cause systemic vasoconstriction.[15] Continuous infusion of ET-1 or ET-3 over 7 days leads to sustained hypertension in rats, mediated through an increase in total peripheral resistance,[16] that can be prevented by salt restriction.[17] The effect of salt restriction may reflect increased vascular sensitivity to ET-1 with a high salt diet or an interaction with the renin-angiotensin system; urinary sodium balance does not change during infusion of ET-1 in these studies.[17]

Studies using [^{125}I]ET-1 show that the sustained increase in arterial pressure occurs despite a rapid clearance of ET-1 from the circulation.[18,19] Over 60% of ET is removed during the first minute (Fig. 6.2), the majority of which binds to the lung, followed by the kidney and liver. These results would be consistent with an extremely slow dissociation of ET-1 from its receptors on vascular smooth muscle, as has been reported in vitro.[20] Bilateral nephrectomy reduces the clearance of exogenous ET-1 and leads to a more prolonged pressor response to ET-1,[21] suggesting that the kidneys play an important role in the clearance of biologically active peptide.

At least two subtypes of vascular ET receptors exist; the ET_A and ET_B receptors (see chapters 3 and 4). Vasoconstriction to ET-1 was initially thought to be mediated solely by vascular smooth muscle cell ET_A receptors, with ET_B receptors restricted to the endothelium, where they mediated generation of endothelium-derived dilator substances. However, the existence of a vasoconstrictor ET_B receptor has been demonstrated subsequently in vitro (see chapters 3 and 4) and selective ET_B receptor agonists have been shown to increase blood pressure in many animal species (Fig. 6.3).[22-26] Further in vivo evidence for a constrictor ET_B receptor has been derived from antagonist studies, in

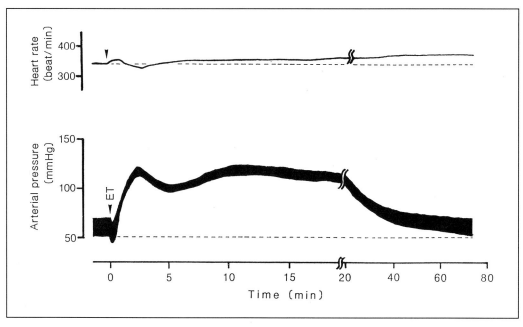

Fig. 6.1. Sustained pressor effect of intravenous bolus injection of porcine ET-1 (ET; 1 nmol/kg) in vivo in the anesthetized, chemically denervated rat. The baseline diastolic pressure is shown by a broken line. Reproduced with permission from Yanagisawa et al, Nature 1988; 332: 411-415.

Fig. 6.2. A typical plasma disappearance curve of total radioactivity after intravenous injection of 30 pmol of [125I]ET-1 together with 1,500 pmol/kg of native ET-3 (upper) and a typical pressor response to intravenous injection of 1,500 pmol/kg of ET-1 (lower) to the anesthetized rat. Radiolabeled ET-1 is rapidly cleared from plasma, in contrast to the sustained pressor action of the peptide. Reproduced with permission from Shiba et al, J Cardiovasc Pharmacol 1989; 13 (Suppl 5): S98-S101.

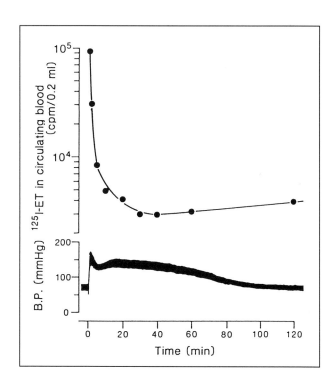

which ET_A receptor antagonists do not fully block the pressor response
to ET-1 (Fig. 6.4).[25,27-30] These results suggest that both ET_A and ET_B
receptors mediate the pressor effects of ET-1. The combined $ET_{A/B}$
receptor antagonist, bosentan, effectively blocks all of the hemodynamic
effects of ET-1, -2 and -3, and of big ET-1,[31] suggesting that non-
ET_A, non-ET_B receptors play little or no role in this response.

The pressor response to ET-1 may be mediated indirectly, through
generation of other constrictor substances. Vasoconstrictor prostanoids
appear to potentiate regional vasoconstriction to ET-1,[32-34] although
cyclo-oxygenation inhibition does not appear to attenuate the pressor
effect to ET-1.[32] The role of the endothelium-derived vasoconstrictor
substance, platelet activating factor, remains unclear, because, although
selective platelet-activating factor receptor antagonists attenuate the
constrictor response to ET in anesthetized rats,[35] they are without effect in conscious rats.[36] The sympathetic and renin-angiotensin systems may contribute to the pressor response; the interactions between
ET-1 and these systems are discussed later.

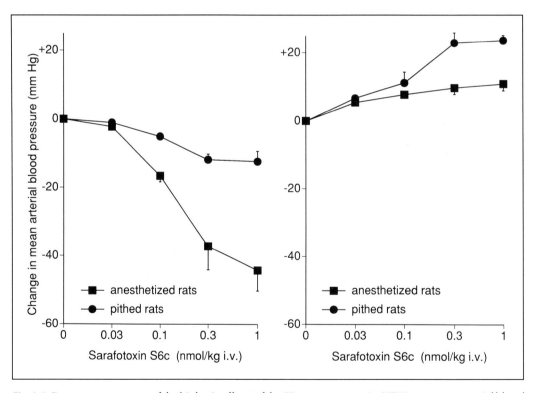

Fig. 6.3. *Dose-response curve of the biphasic effects of the ET_B receptor agonist SFTX6c on mean arterial blood pressure in anesthetized (n = 4) and pithed rats (n = 4). The depressor response (left) always preceded the pressor response (right). These findings suggest the presence of both vasodilator and vasoconstrictor ET_B receptors. Reproduced with permission from Clozel et al, Biochem Biophys Res Commun 1992; 186: 867-873.*

Intravenous infusion of ET-1 has been shown to increase blood pressure in humans.[37-42] As in animals the effect is slow in onset. This pressor response is not affected by pretreatment with cyclosporin, the Ca^{2+} antagonist nifedipine, or cyclo-oxygenase inhibition with indomethacin.[40]

DEPRESSOR RESPONSES

All three ET peptides elicit a transient hypotensive response after bolus doses, lasting only a few minutes, which precedes the sustained pressor effect and is most marked for ET-3.[5] The selective ET_B receptor agonists, SFTX S6c and [Ala 1,3,11,15] ET-1 cause transient hypotension in rats (Fig. 6.3) and cynomolgus monkeys.[26] From studies using analogues of sarafotoxin in rats, it appears that the N-terminal amino group of the molecule determines the depressor response.[43] The hypotension may reflect a pharmacological, rather than physiological,

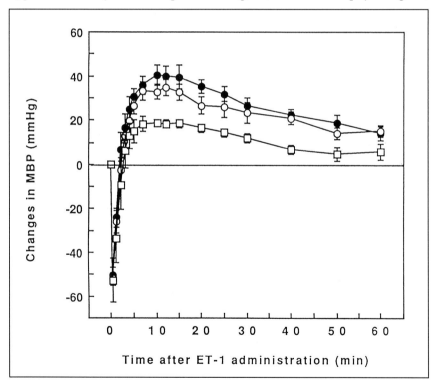

Fig. 6.4. Effect of the ET_A receptor antagonist, BE-18257B, on depressor and pressor responses to intravenous ET-1 (1 nmol/kg) in three groups of four conscious and unrestrained Wistar Kyoto rats. BE-18257B was administered intraperitoneally at 10mg/kg (O) or 50mg/kg (□) 60 minutes before ET-1; control animals (●). The ET_A selectivity of this receptor antagonist is shown by its lack of effect on the ET_B receptor-mediated transient depressor response. The incomplete abolition of the pressor response supports the presence of non-ET_A receptors mediating vasoconstriction. Reproduced with permission from Ihara et al, Biochem Biophys Res Commun 1991; 178: 132-137.

response to the briefly sustained, high concentrations of ETs that occur following bolus administration.[44] Under more physiological conditions, in which ET concentrations rise more slowly, such as occur following intravenous infusion of ETs[45] or big ETs,[46] hypotension does not occur. Nevertheless, the hypotensive response to bolus administration is useful to demonstrate the endothelial actions of the ET family.

The depressor response is thought to be caused by generation of endothelium-dependent dilating factors, primarily nitric oxide and prostacyclin, mediated by endothelial cell ET_B receptors. Inhibitors of nitric oxide synthase[47-54] appear to attenuate the hypotensive response to ET-1, although the attenuation is often incomplete,[48,50,53] implying that other factors may contribute. Inhibition of cyclo-oxygenase[52,54,55] also reduces the depressor effect, although this is not a universal finding.[56] In addition, the sustained pressor effect of ET is usually potentiated by inhibition of nitric oxide[48,53] and prostaglandin generation,[52,56] indicating that the pressor effect of ET is limited by the stimulated release of endothelium-dependent vasodilators. Furthermore, the ability of ET_A receptor antagonists to enhance the vasodilator activity of ET also suggests a functional antagonism between constrictor and dilator actions.[57] However, some investigators have found no major role for nitric oxide or prostacyclin in mediating the depressor response,[44,58] suggesting that other mediators may be involved. Endothelium-derived hyperpolarizing factor is a candidate, but its contribution to ET-1 mediated vasodilatation has only been examined in the lung (see Pulmonary Section). ET-1-mediated generation of atrial natriuretic peptide does not appear to account for the depressor response.[59]

Depressor responses to ET-1 have not been shown in humans, probably because ET-1 has been administered by intravenous infusion rather than as bolus doses. Transient regional vasodilator responses have been observed and are discussed later.

REGIONAL PHARMACOLOGY

The hemodynamic effects of systemically administered ETs are the cumulative sum of their actions on the heart, peripheral vasculature, nervous system, kidney and endocrine glands. There appear to be wide regional variations in the responses to ETs in vivo. Furthermore, the dose and mode of administration appear to influence the actions of the ET family in different vascular beds. It is likely that the regional variation in responses can be accounted for by differences in the balance between constrictor and dilator responses to the ETs, probably as a result of varying densities of vasoconstrictor smooth muscle and vasodilator endothelial cell receptors.

CARDIAC

ET-1 is a potent and long-lasting constrictor of coronary vessels in vivo in the dog,[60-62] rabbit,[63] monkey,[64] goat[13] and pig,[12,65] resulting in a marked fall in coronary blood flow and myocardial ischaemia. Coronary angiograms reveal these effects are primarily due to the contraction of small coronary arteries.[60,63] The coronary resistance bed is more sensitive to the effects of ET than other resistance beds, excepting the kidney.[64]

The in vivo actions of ET on the heart are unclear, even though positive chronotropic and inotropic effects have been found in vitro. Several studies find the initial hypotension produced by ET is associated with an increase in heart rate and cardiac output, suggesting that it is due to systemic vasodilatation.

In contrast, the pressor response to high doses of ET is associated with bradycardia and a fall in stroke volume, causing a marked reduction in cardiac index.[61,66] The chronotropic effects of ET-1 in vivo appear to be reflex in origin, because heart rate does not change when ET-1 is administered after blockade of cardiac efferent neural mechanisms.[67] The reduction in stroke volume is probably due to a combination of systemic vasoconstriction increasing afterload, and coronary vasoconstriction causing myocardial ischaemia.[61,66] Pressor doses of ET-1 have been shown to be positively inotropic in conscious rats.[67,68]

ET has also been shown to cause fatal ventricular arrhythmias,[65] possibly via a distinct proarrhythmic effect separate from myocardial ischemia caused by coronary vasoconstriction.

In humans, infusion of ET-1 tends to decrease cardiac output,[39] probably through a baroreceptor mediated decrease in heart rate, although an increase in afterload may also contribute. Coronary vasoconstriction has not been observed in clinical studies, but does occur in humans who are bitten by the burrowing asp, *Actractaspis engaddensis*, whose venom contains SFTXs.[69]

SKELETAL MUSCLE, SKIN AND MESENTERIC VASCULAR BEDS

Hindquarter blood flow increases during the initial hypotensive phase after bolus administration of ET-1 to rats.[58,70] Nitric oxide does not seem to play a major role in the hindquarters vasodilator response to ET because N^G-monomethyl-L-arginine, an inhibitor of the nitric oxide synthase enzyme, does not significantly attenuate the hindquarters vasodilatation to ET.[58] During the pressor phase, or during intravenous infusion where there is no initial fall in blood pressure, no change in hindquarters blood flow occurs, although there are marked decreases in blood flow to other vascular beds. ET may be more effective in stimulating vasodilator mechanisms in the hindquarter bed, perhaps due to a higher density of dilator ET_B receptors present on endothelial cells or a reduced number of constrictor vascular smooth muscle $ET_{A/B}$ receptors, compared to other vascular beds. There may be species differences, because infusion of ET-1 locally into the femoral artery of the anesthetized dog results in a dose-dependent reduction in hindlimb blood flow, without vasodilatation.[11]

Although ET-1 can cause mesenteric vasodilatation in vitro,[56] this is seldom observed in vivo, the predominant response being vasoconstriction.[32] Systemically administered ET-1 causes more vasoconstriction in the mesenteric bed than in the hindquarters skeletal muscle bed.[7,58,71,72] Both ET-1 and the selective ET_B receptor agonist SFTX S6c cause mesenteric vasoconstriction in vivo, suggesting that ET_A and ET_B receptors mediate this response.[24]

In humans, brachial artery infusion of ET-1 slowly decreases forearm blood flow in a dose-dependent manner,[73-81] with the effect sustained for more than two hours after halting the infusion.[73] Co-infusion of dihydropyridine Ca^{2+} channel antagonists blocks forearm vasoconstriction,[73,78-80] but this blockade is not specific to ET-1 because vasoconstriction to angiotensin II is blocked to an equal or greater degree.[73] Thus, at least some of the effect of Ca^{2+} channel antagonists are likely to be mediated against basal arteriolar tone, rather than specifically against vasoconstriction to ET-1. Furthermore, dorsal hand vein constriction to ET-1 is blocked more effectively by an opener of ATP-sensitive K^+ channels than a Ca^{2+} antagonist,[82] suggesting that ET-1 responses in human veins depend only in part on Ca^{2+} entry through dihydropyridine-sensitive Ca^{2+} channels. In addition, the greater efficacy of K^+ channel opening agents is consistent with ET-1 acting to close ATP-sensitive K^+ channels, thus leading to plasma membrane depolarization and vasoconstriction by mechanisms additional to opening of voltage-operated Ca^{2+} channels.

Both ET_A and ET_B receptors appear to mediate vasoconstriction to ET-1 in humans, because selective ET_B receptor agonists, such as ET-3 and SFTX S6c, constrict human resistance and capacitance vessels in vivo.[83] There may be regional differences in the ET_B receptor mediated vasoconstriction because vasoconstriction to ET-1 in the skin microcirculation appears to be mediated solely by ET_A receptors.[84] ET-3, but not ET-1, causes transient forearm vasodilatation when administered as a bolus via the brachial artery, consistent with the involvement of an endothelial ET_B receptor.[83] However, venodilatation to the ETs has not been shown in human skin capacitance vessels.[85] Nonetheless, inhibition of prostaglandin, but not nitric oxide, generation potentiates venoconstriction to ET-1 and SFTX S6c in vivo in man.[86,87] Thus, the venous endothelium may generate vasodilator substances in response to ET-1, but the vasodilator effects of such substances appear to be masked by the simultaneous direct venoconstriction caused by the peptide, and serve only to modulate venoconstriction.

RENAL

ET-1 has two main direct actions on the kidney, causing renal vasoconstriction and tubular sodium and water loss, these actions probably reflecting separate sites of production in blood vessels and renal tubules.

ET-1 is a potent constrictor of both afferent and efferent glomerular arterioles in vivo, resulting in reduced renal plasma flow and glomerular filtration rate, and hence reduced urine flow and sodium excretion.[15,61,88,89] The renal vasoconstrictor action of ET-1 appears to be mediated by different receptor subtypes in different species. Studies using selective ET_A receptor antagonists and ET_B receptor agonists, show that ET-1-induced renal vasoconstriction involves ET_A receptors in the dog,[90] ET_B receptors in the rat,[91,92] and both receptor subtypes in the pig.[57,93] The involvement of prostanoids in the renal effects has been studied, but results are conflicting. In one study, cyclo-oxygenase inhibitors had no effect on the renal actions of ET-1 in rats.[94] In contrast, in another study in rats, co-administration of indomethacin and ET-1 causes selective relaxation of afferent arterioles, preserving glomerular filtration rate, thus supporting a role for prostanoids in ET-1-induced arteriolar constriction.[95]

ET-1 also has diuretic and natriuretic effects.[88,96] Whereas King et al propose these effects are partly pressure related,[88] Perico et al reported that the natriuresis and diuresis were independent of any renal hemodynamic changes, because they occurred at a dose of ET-1 insufficient to affect renal blood flow.[96] Both groups suggest that a direct action of ET-1 on the proximal tubules contributes to the diuretic and natriuretic effect. This appears to be mediated by tubular ET_B receptors[97] and probably reflects the action of endogenously generated ET-1 within renal tubules. Atrial natriuretic peptide is unlikely to mediate ET-1-induced natriuresis because plasma levels remain unchanged after ET-1 infusion.[96] In addition, Perico and colleagues reported that oral administration of the specific 5-lipoxygenase inhibitor, L-651,392, and the specific leukotriene C4/D4 receptor antagonist, L-649,923, prevented the diuretic and natriuretic effect of ET-1,[96] suggesting a role for 5-lipoxygenase products in the renal tubular effects of ET-1.

In humans, systemic infusion of ET-1 causes renal vasoconstriction, similar to that observed in the splanchnic bed[38] but more marked than that observed in the leg.[41] Infusion of ET-1 decreases renal sodium excretion,[42] even at doses insufficient to cause renal vasoconstriction.[98] Thus there is no in vivo evidence in humans for a renal tubular diuretic action of ET-1. Such an effect is most likely to be detected using selective ET receptor antagonists.

PULMONARY

ET-1 is a potent dilator, at low doses,[99-101] and constrictor at high doses,[102,103] of the pulmonary vascular bed in vivo; the balance between these responses depends on the dose and underlying tone of the preparation used. The pulmonary vasodilator response to ET is mediated in part by the release of nitric oxide, because inhibition of nitric oxide synthase blocks pulmonary vasodilatation.[101] Similarly, because glibenclamide, an inhibitor of ATP-dependent potassium channels, inhibits the dilatation,[99,101] endothelium-derived hyperpolarizing factor may mediate this response. Prostacyclin is unlikely to be involved because cyclo-oxygenase inhibitors do not affect dilatation to ET-1.[101] The pulmonary vasoconstrictor action of ET appears to be partly mediated by cyclo-oxygenase products, especially thromboxane A_2 in the rat and guinea pig, but not in the cat or rabbit.[104] As in the systemic circulation, both ET_A and ET_B receptors mediate pulmonary vasoconstriction.

Only one study has examined the effects of ET-1 on pulmonary vascular resistance in humans.[39] Here, intravenous infusion of a relatively low dose of ET-1 did not increase pulmonary vascular resistance, although the dose was sufficient to increase systemic vascular resistance by about 10%. The contrast between this result and data from animals may be related to differences in the dose and mode of administration of ET-1 used in different studies.

The actions of ET-1 on the airways include bronchoconstriction (ET_B mediated), mucus secretion (ET_B mediated) and bronchial smooth muscle growth (ET_A mediated); these actions have been reviewed in depth elsewhere.[1]

CENTRAL AND PERIPHERAL NERVOUS SYSTEM

The ETs are potent constrictors of cerebral arteries in vivo,[13,105,106] and the effects of ET-1 on the basilar artery of the dog last for at least three days after intracisternal injection.[105] ET-1 has central actions that may contribute to its pressor actions. Intracerebroventricular administration of low doses of ET-1 increases blood pressure through stimulation of central sympathetic outflow in the rat[107,108] and rabbit.[109] This pressor effect occurs at doses of ET-1 that are not sufficient to increase blood pressure when administered intravenously. In addition, centrally administered ET-1 appears to sensitize the baroreceptor reflex by reducing resting parasympathetic tone.[110] Furthermore, chronic intracerebroventricular infusion of low dose ET-1 for seven days slowly increases blood pressure, in association with increased urinary catecholamine and vasopressin excretion.[111] Finally, topical administration of ET-3 to the hypothalamus exerts profound effects on salt and water balance in rats, with substantial inhibition of thirst in water deprived animals,[112] suggesting that ET is involved in fluid and electrolyte regulation.

ET-1 may also have a role in the peripheral autonomic nervous system. Binding sites for ET-1 are present in the carotid bifurcation, and topical application of the peptide inhibits baroreceptor, and stimulates chemoreceptor, responses at this site.[113] In addition, ET-1 may potentiate the peripheral actions of the sympathetic nervous system in threshold doses.[114,115] Furthermore, ET-1 markedly elevates venous tone in vivo in rats via a reflex increase in sympathetic nerve activity and activation of alpha adrenoreceptors.[116] However, in man, a peripheral interaction with the sympathetic nervous system has only been demonstrated in patients with essential hypertension.[77,117]

The ETs have been postulated to have other effects in the nervous system including involvement in pain pathways; these areas have been reviewed in detail elsewhere.[2]

ENDOCRINE SYSTEM

Although ET-1 inhibits release of renin from isolated rat glomeruli,[118] its other actions tend to increase the activity of the renin-angiotensin-aldosterone system. ET-1 stimulates endothelial angiotensin converting enzyme activity,[119] and stimulates the tissue renin-angiotensin system of the rat mesenteric bed, increasing generation of renin and angiotensin II.[120] In the adrenal gland, ET-1 stimulates release of aldosterone from isolated cortical zona glomerulosa cells,[121] and adrenaline from medullary chromaffin cells.[122] However, in vivo administration of ET-1 to animals increases renin, aldosterone and adrenaline concentrations;[123,124] renin concentration apparently rises because of ET-1 induced renal vasoconstriction. The acute pressor response to ET-1 is unaffected by pretreatment with the angiotensin converting enzyme inhibitor captopril.[124] However, renal vasoconstriction to systemic ET-1 is abolished by captopril, implying some regional variation in the interaction between ET-1 and the renin-angiotensin system. In contrast, hypertension caused by chronic infusion of ET-1 can be completely prevented by concomitant administration of captopril, suggesting that the chronic pressor response may be mediated via the renin-angiotensin system.[125] However, chronic infusion of ET-1 does not increase plasma angiotensin II concentrations,[125] implying either that ET-1 interacts with a tissue, not circulating, renin-angiotensin system, or that the effect of captopril is not mediated through decreased angiotensin II generation, but perhaps through an effect on kinin breakdown.

Concentrations of both ET-1 and vasopressin rise in parallel during upright tilt in normal subjects and are unchanged during tilt in patients with diabetes insipidus.[126] Although the suggestion has been made that ET-1 is acting as a classical circulating hormone in these circumstances, this seems unlikely, as its concentration is below that necessary to cause direct vasoconstriction. In addition, it is likely that such rapid changes in ET concentrations reflect alterations in clearance rather than generation. However, ET-1 may have an autocrine or paracrine role in control of vasopressin release as infusion of the peptide increases vasopressin concentrations in dogs.[123] Circulating concentrations of atrial natriuretic peptide are increased by infusion of ET-1 in rats,[68,127] and pretreatment of rats with antiserum to atrial natriuretic peptide potentiates the pressor response to ET-1.[128] Thus, endogenously generated atrial natriuretic peptide may modulate the vasoconstrictor actions of ET-1 in vivo. Brain natriuretic peptide concentrations are increased after bolus injection of ET-1 in rats.[129] The release of brain natriuretic peptide appears to be stimulated partly through a change in blood pressure, and partly through direct stimulation by ET-1.

The other endocrine actions of ET have been reviewed in detail elsewhere.[3]

ECE AND ET GENERATION

ET-1 is generated through the action of an 'ET converting enzyme' (ECE) on a 38 amino acid precursor, big ET-1 (see chapter 2). Bolus intravenous administration of big ET-1 causes a pressor effect of similar magnitude to that of ET-1 in conscious,[130-133] and anesthetized rats,[134,135] guinea pigs,[136] and rabbits.[137] In the anesthetized pig, however, porcine big ET-1 has poor vasopressor activity compared to ET-1.[138] This result may be due to the mode of administration of big ET-1, because the other studies cited gave bolus injections, while Hemsén et al[138] gave infusions, or it may be due to species differences.

Big ET-1 is ~100 times less potent as a direct vasoconstrictor of blood vessels in vitro than ET-1,[130] suggesting that big ET-1 does not exert its actions through direct binding to ET receptors. Confirmation that the pressor activity of big ET-1 is dependent on cleavage to ET-1 through ECE comes from the inhibition by the ECE inhibitor phosphoramidon of the pressor response to big ET-1 (Fig. 6.5).[131-136] Phosphoramidon did not inhibit pressor responses to ET-1 in these studies. Inhibitors of angiotensin converting enzyme and neutral endopeptidase do not block the pressor effects of big ET-1.[133,135] There may be regional variation in ECE activity, because mesenteric vasoconstriction to systemic big ET-1 is less sensitive to phosphoramidon than vasoconstriction in the renal and hindquarters vascular beds.[131] This result also suggests conversion of big ET-1 occurs locally rather than in the systemic circulation.

Big ET-1 has opposite effects to ET-1 on renal function in conscious rats, in doses with similar pressor effects, causing diuresis and natriuresis while ET-1 markedly decreases these parameters. In addition, big ET-1 causes smaller decreases in renal blood flow and glomerular filtration rate than ET-1.[46,89] This difference may be because big ET-1 is more readily converted to the mature peptide in tubular capillaries, where ET-1 mediates diuresis and natriuresis, than in the glomeruli and arterioles, where ET-1 decreases renal blood flow and glomerular filtration rate.

The precursors of ET-2 and ET-3 also have vasoconstrictor actions similar to their mature forms when administered in vivo to conscious[139] and anesthetized rats,[140,141] although the magnitude of the response is smaller than that of ET-2 and ET-3. Phosphoramidon can inhibit the actions of big ET-2 and big ET-3 in rats.[139-141] In contrast to studies in the rat, D'Orleans-Juste and colleagues found big ET-3 had no pressor response in the anesthetized guinea pig.[137] The authors suggest ECE is selective for the N- and C-terminal sequences of big ET-1. These varying results with big ET-2 and big ET-3, together

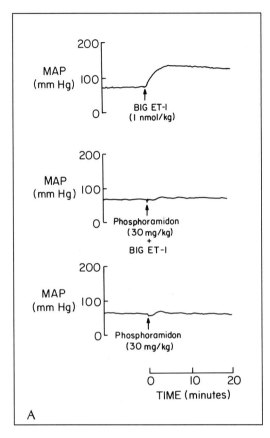

Fig. 6.5. (A) Representative tracings showing the effect on mean
arterial pressure (MAP) of a bolus
intravenous injection of big ET-1
at 1 nmol/kg (top panel), phosphoramidon at 30mg/kg injected
just prior to big ET (1nmol/kg)
(middle panel), and phosphoramidon at 30mg/kg administered
alone (bottom panel) in a ganglion-blocked anesthetized rat.

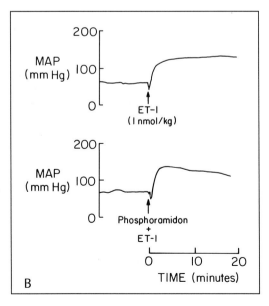

(B) Representative tracings showing the effect on MAP of a bolus
intravenous injection of ET-1 at 1
nmol/kg (top panel) and the effects of phosphoramidon at 30
mg/kg injected just prior to ET-1
(1 nmol/kg) (bottom panel) in a
ganglion-blocked anesthetized rat.
Phosphoramidon blocks the pressor response to big ET-1 but not to
ET-1, demonstrating its action as
an ECE inhibitor. Reproduced with
permission from McMahon et al,
J Cardiovasc Pharmacol 1991; 17
(Suppl 7): S29-S33.

with the recent in vitro isolation and characterization of an ECE (ECE-1) selective for big ET-1 (see chapter 2), suggest the presence of more than one ECE. The contrasting findings in vivo may reflect species variation in the expressions of different ECE isoforms.

In humans, brachial artery administration of big ET-1 causes a dose-dependent forearm vasoconstriction that is completely blocked by phosphoramidon (Fig. 6.8), suggesting that the effects of the precursor are mediated through conversion to the mature peptide by ECE.[81] The blockade of constriction to big ET-1 by phosphoramidon is unlikely to have been due to inhibition of ET receptor binding, because vasoconstriction to ET-1 was unaffected by phosphoramidon. Because circulating blood does not exhibit ECE activity,[142] conversion of big ET-1 in the forearm probably occurs via vascular ECE situated within the forearm blood vessels. Big ET-1 is approximately 10 fold less potent than ET-1, suggesting that local ECE converts about 10% of luminally presented big ET-1 to ET-1.

Further evidence that generation of mature ET-1 is responsible for forearm vasoconstriction to big ET-1, through the action of a phosphoramidon-sensitive ECE, is derived from measurement of the concentrations of ET peptides in blood drawn from antecubital veins draining infused and noninfused forearms. Plasma immunoreactive ET-1, big ET-1, and the inactive C-terminal fragment of ET-1 (CTF) that is formed during the cleavage of big ET-1 by ECE, have been measured in these studies and reported in a preliminary form.[143] Concentrations of ET-1, big ET-1, and CTF in venous plasma from the infused arm, but not the control arm, increase significantly during infusion of big ET-1. The ratio of CTF to big ET-1 is about 0.1:1, indicating ~10% conversion of the precursor by ECE, in keeping with the functional results. There is no significant increase in plasma ET-1 when big ET-1 is co-infused with phosphoramidon.

Xu and colleagues recently reported the isolation, structure and characterization of ECE-1, a novel membrane-bound neutral metalloprotease that is expressed abundantly in endothelial cells in vivo, and that is structurally related to neutral endopeptidase 24.11 (see chapter 2).[144] Interestingly, compared with 50-90% conversion of endogenous big ET-1 in cells transfected with cDNA for ECE-1, the conversion of exogenous big ET-1 by such cells was much less efficient [~10%] and similar to that found in the functional and biochemical studies reviewed above.

PHYSIOLOGICAL ROLE OF ET-1 IN MAINTENANCE OF VASCULAR TONE

TRANSGENIC STUDIES

There has been much interest in whether endogenous generation of ET plays a physiological role in the maintenance of basal vascular tone and blood pressure. One novel approach to this issue has been to examine the hemodynamics of transgenic animal models which have had additional ET gene(s) inserted or the genes deleted. It has not proved possible to develop a transgenic model with an extra ET-1 gene, because such fetuses are not viable, suggesting that ET-1 plays a fundamental role in fetal development.[145] Addition of an extra ET-2 gene does not appear to alter blood pressure.[146] This may be because ET-2 in this model is primarily expressed not in the systemic vasculature, but in the kidney, where any vasoconstrictor action to increase blood pressure will be opposed by tubular natriuretic and diuretic actions. In addition, the ETs are locally active substances, and although circulating plasma concentrations of ET-2 in the transgenic rats were increased ~5 fold, such concentrations would not usually be sufficient to cause systemic vasoconstriction. Also, because there is 15 fold increase in circulating big ET-2 concentrations, there appears to be down-regulation of ECE mediated cleavage of big ET-2 in the transgenic animals.[146]

Transgenic mice lines with ET-1 genes deleted have also been established. In mice with both ET-1 genes deleted there are serious craniofacial abnormalities resulting in death from asphyxia soon after birth.[147] Mice with one ET-1 gene deleted paradoxically have slightly higher blood pressure than controls, despite lower circulating concentrations of ET-1.[147] However, given the apparently crucial role of ET-1 in fetal craniofacial development, this elevation of blood pressure could be due to hypoxia from sub-clinical upper airway deformities. In addition, such 'knockout' models are not designed to address the physiological contribution of ET-1 to cardiovascular regulation in the fully developed adult, but may give new information about their role in development.

ECE INHIBITOR AND RECEPTOR ANTAGONIST STUDIES IN ANIMALS

In contrast to the results from transgenic animals, there is increasing evidence from studies using ECE inhibitors and ET receptor antagonists that basal generation of ET-1 contributes to the maintenance of basal vascular tone and to regulation of blood pressure. The best evidence has come from carefully designed, adequately powered studies that have followed hemodynamic responses for several hours after administration of an ECE inhibitor or ET receptor antagonist. Such studies are able to take into account the slow reversal of ET-1-induced vasoconstriction by anti-ET agents.[30] Inhibition of ET generation by administration of the ECE inhibitor phosphoramidon slowly decreases mean arterial pressure in normotensive and spontaneously hypertensive rats over several hours (Fig. 6.6).[133] Similar results have since been obtained using ET receptor antagonists. Intravenous infusion of the selective ET_A receptor antagonist, BQ-123, into normo-

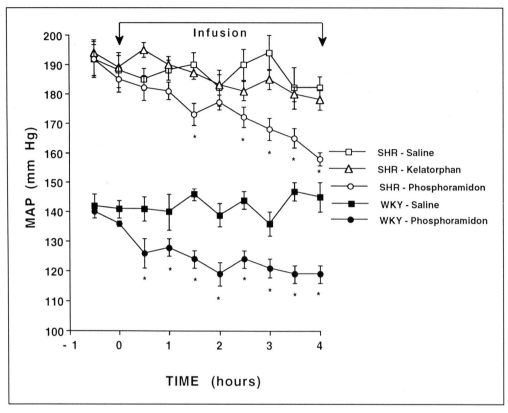

Fig. 6.6. Effect of intravenous infusion of saline (3 ml/h), the ET converting enzyme and neutral endopeptidase inhibitor phosphoramidon (20 mg/kg/h), and the neutral endopeptidase inhibitor kelatorphan (20 mg/kg/h) on mean arterial pressure (MAP) in conscious spontaneously hypertensive (SHR) and Wistar Kyoto (WKY) rats (n=3-6). *p < 0.05 compared to saline-infused SHRs or WKY rats. Phosphoramidon, but not kelatorphan, lowers blood pressure by ~15% in both groups of rats, suggesting a role for endogenous generation of ET-1 in the physiological maintenance of blood pressure. Reproduced with permission from McMahon et al, J Cardiovasc Pharmacol 1991; 17 (Suppl 7): S29-S33.

tensive Sprague-Dawley rats, decreases mean arterial pressure by ~15% over 60 minutes.[92] In addition, a six hour infusion of BQ-123 decreases blood pressure by ~15% in Wistar-Kyoto normotensive and spontaneously hypertensive rats.[148] Although blood pressure did not decrease significantly in normotensive rats in this study, this was probably owing to the very small numbers studied, as well as the difficulty in detecting the smaller, absolute changes in blood pressure in such animals. The statistically significant correlation between basal blood pressure and fall in blood pressure observed in this study in all hypertensive and normotensive animals studied (Fig. 6.7), suggests that the relative magnitude of the hypotensive response to the ET_A antagonist is no greater in hypertensive than normotensive rats. Similar findings were

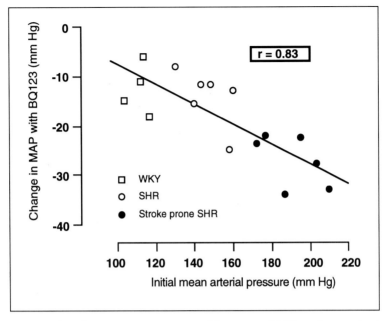

Fig. 6.7. Plot of pretreatment mean arterial pressure (MAP) against maximal changes in MAP after administration of the ETA receptor antagonist BQ-123 at 30 mg/kg/hr for 6 hours in Wistar-Kyoto rats (WKY), spontaneously hypertensive rats (SHR) and stroke prone rats SHR. BQ-123 decreases blood pressure, with the magnitude of the blood pressure fall related to the pretreatment MAP. This supports a phsiological role for endogenously generated ET-1 in the maintenance of blood pressure, as opposed to a specific pathophysiological role in the elevated blood pressure of hypertension. Adapted from Nishikibe et al, Life Sci 1993; 52:717-724 with kind permission from Elsevier Science Ltd, The Boulevard, Langford Lane, Kidlington 0X5 1GB, UK.

also apparent in another study that examined responses to BQ-123 in normotensive and hypertensive rats.[149] Furthermore, the combined $ET_{A/B}$ antagonist Ro-46-2005 has been shown to dose-dependently lower blood pressure by ~30% over several hours in sodium-depleted squirrel monkeys.[150] Finally, the ET_A antagonist, BQ-123, and the $ET_{A/B}$ antagonist, bosentan, both lower blood pressure by ~20% in anesthetized and conscious guinea pigs.[151] The hypotensive effect of bosentan is not affected by blockade of the sympathetic or parasympathetic nervous systems, or by inhibition of renin, cyclo-oxygenase, bradykinin or nitric oxide. Bosentan has no additional hypotensive effect to BQ-123 when co-administered with the ET_A antagonist, suggesting that ET-1 contributes to the maintenance of resting arterial blood pressure through activation of ET_A receptors in these animals. Together these results imply that ET-1 contributes to the maintenance of blood pressure under physiological conditions, at least in some species, probably through its actions as a potent arteriolar vasoconstrictor.

ECE Inhibitor and Receptor Antagonist Studies in Humans

To date, only one study has specifically examined this issue in humans.[81] Brachial artery infusion of the ECE inhibitor phosphoramidon causes progressive vasodilatation of forearm resistance vessels, suggesting that generation of ET-1 by ECE provides an important contribution to the maintenance of basal vascular tone in man (Fig. 6.8). However, phosphoramidon also inhibits neutral endopeptidase, an enzyme that degrades a number of peptide hormones including the natriuretic peptides, angiotensins and ETs.[152,153] Vasodilatation to phosphoramidon could therefore be due to inhibition of forearm neutral endopeptidase causing local accumulation of dilator natriuretic peptides. This possibility appears highly unlikely because brachial artery infusion of the selective neutral endopeptidase inhibitor, thiorphan, causes forearm vasoconstriction rather than vasodilatation. Thus, the forearm vasodilator effects of phosphoramidon through ECE inhibition may be offset to some degree by vasoconstriction through inhibition of neutral endopeptidase causing impaired breakdown of constrictor peptides, such as angiotensin II and ET-1.[152,153] Such vasoconstriction may explain why systemic inhibition of neutral endopeptidase fails to decrease blood pressure despite increasing sodium excretion.[154]

There appear to be two cellular sites for ECE, with low efficiency cell surface conversion of exogenous big ET-1 and high efficiency processing of endogenous big ET-1 in membrane associated organelles.[144] Given that phosphoramidon is less potent as an inhibitor of ECE-1 mediated cleavage of endogenous big ET-1, it is possible that the vasodilatation observed with phosphoramidon infusion was not maximal, even though the dose was sufficient to block conversion of exogenous big ET-1. This contention is supported by the results of human studies with BQ-123, the specific and selective peptide ET_A receptor antagonist.

Brachial artery infusion of BQ-123 prevents vasoconstriction to locally administered ET-1, even 30 minutes after halting BQ-123 (Fig. 6.9). Infusion of BQ-123 alone causes progressive and substantial forearm vasodilatation, greater than that observed to phosphoramidon. These results suggest that endogenous generation of ET-1 maintains vascular tone in man through activation of ET_A receptors. However, a role for ET_B receptors in mediating vasoconstriction to endogenously generated ET-1 cannot presently be excluded, given that exogenous ET-1 can produce vasoconstriction through an action on ET_B receptors.[83]

The only other factors, besides ET-1, that have been shown to have a similar fundamental physiological role in maintenance of basal vascular tone are the sympathetic nervous system and nitric oxide.[155] Sustained vasoconstriction elicited by ET-1 may act in concert with the short lived effects of the sympathetic nervous system and nitric oxide to stabilize vasomotor tone, while preserving flexibility in its

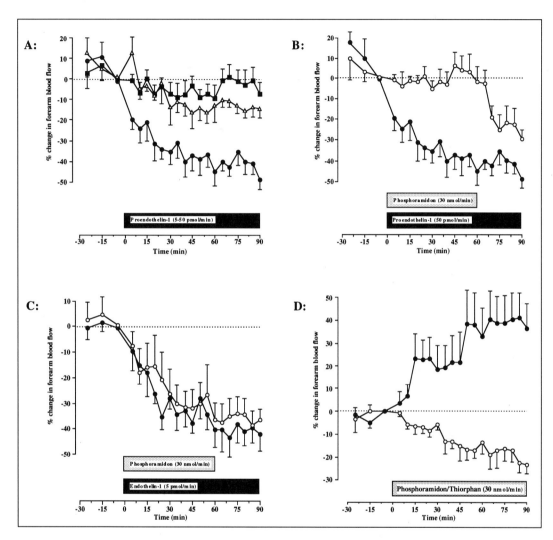

Fig. 6.8. (A) Brachial artery infusion of big ET-1 at 5 (■), 15 (Δ), and 50 pmol/min (●) causes slow-onset forearm vasoconstriction in healthy men, demonstrating the presence of ECE in human blood vessels in vivo. (B) Forearm vasoconstriction to intra-arterial big ET-1 (●; 50 pmol/min) is abolished by co-infusion of the ECE and neutral endopeptidase inhibitor phosphoramidon (O; 30 nmol/min). (C) Forearm vasoconstriction to intra-arterial ET-1 (●; 5 pmol/min) is unaffected by co-infusion of the ECE inhibitor phosphoramidon (O; 30 nmol/min), demonstrating that the effect of phosphoramidon on big ET-1 responses is not due to inhibition of responses to the mature peptide. (D) Brachial artery infusion of the ECE and neutral endopeptidase inhibitor phosphoramidon (●; 30 nmol/min) results in slow-onset forearm vasodilatation. In contrast, the more selective neutral endopeptidase inhibitor thiorphan (O; 30 nmol/min) causes modest forearm vasoconstriction. Thus, ECE inhibition with phosphoramidon causes forearm vasodilatation, offset to some degree by vasoconstriction caused by neutral endopeptidase inhibition. Together these results suggest a role for generation of ET-1, through a vascular ECE, in the maintenance of basal vascular tone in man. Values are mean ± SEM; n = 6 for each group. Adapted from WG Haynes, DJ Webb. Lancet 1994; 344: 852-854 with permission.

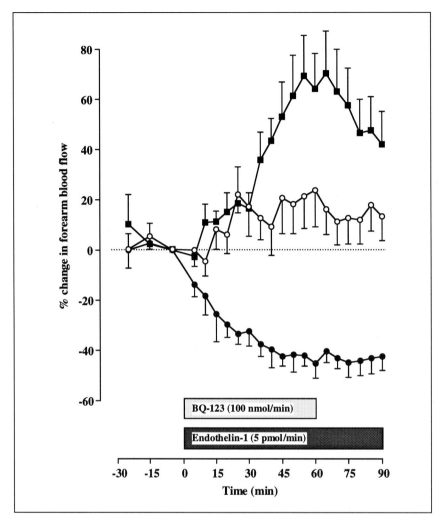

Fig. 6.9. Brachial artery infusion of ET-1 to healthy human subjects causes significant forearm vasoconstriction (●) that can be abolished by co-infusion of the ET_A receptor antagonist, BQ-123 (○). Infusion of BQ-123 alone causes progressive and substantial forearm vasodilatation (■). Values are mean ± SEM; n = 6 for each group. These results support a physiological role for endogenous generation of ET-1, acting through ET_A receptors, in the maintenance of basal vascular tone in healthy man. Reproduced from WG Haynes, DJ Webb. Lancet 1994; 344: 852-854 with permission.

dynamic control. These results, and those from animals reported above, suggest that orally-active ECE inhibitors and ET receptor antagonists may have potential therapeutic uses as novel vasodilator agents.

FUTURE DIRECTIONS

The ETs are potent vasoconstrictor and pressor peptides with uniquely sustained actions. They have indirect effects on the kidney, endocrine glands and central and peripheral nervous system that act to augment their pressor effects. The ETs act indirectly on the endothelium to modulate their direct vasoconstrictor actions by stimulating generation of nitric oxide and prostaglandins. Despite these many actions, doubt had been expressed about the physiological relevance of the ETs. However, recent studies suggest that ET-1 plays a fundamental physiological role in regulation of basal vascular tone in humans. Nonetheless, there still are a number of areas in the cardiovascular physiology and pharmacology of the ETs that await clarification. First, although ET-1 appears to contribute to vascular tone in the forearm bed, we do not know whether ECE inhibitors or ET receptor antagonists will lower blood pressure in healthy humans. This is of some importance because it will undoubtedly influence the clinical development of such drugs. Second, the reasons for the widespread heterogeneity in pharmacological responses to the ETs need to be explored further. This is particularly important in regard to the function and selectivity of ECE in different vascular beds and tissues, and to the function of ET_B receptors. We still do not know if the heterogeneity reflects differences between studies in the size of vessel, type of vessel, regional vascular bed or the species examined, or whether many of the apparent differences are simply due to the experimental methodology used or to random variation, particularly important where small numbers of experimental animals have been examined. Third, although there is evidence from agonist pharmacological studies that ET_B receptors can mediate vasoconstriction, we do not know the physiological role of these receptors. Given their role in mediating endothelium-dependent dilatation, appropriate in vivo studies using selective ET_B antagonists are required to elucidate whether vasoconstriction or vasodilatation is the predominant physiological action of these receptors.

The increasing knowledge gained from other disciplines, such as molecular biology, and the recent availability of highly potent and selective pharmacological tools, will help the future investigation of the cardiovascular physiology and pharmacology of the ETs.

REFERENCES

1. Hay DWP, Henry PJ, Goldie RG. Endothelin and the respiratory system. Trends Pharmacol Sci 1993; 14: 29-32.
2. Stojilkovic SS, Catt KJ. Neuroendocrine actions of the endothelins. Trends Pharmacol Sci 1992; 13: 385-391.
3. Kennedy RL, Haynes WG, Webb DJ. Endothelins as regulators of growth and function in endocrine tissues. Clin Endocrinol 1993; 39: 259-265.
4. Yanagisawa M, Kurihara H, Kimura S et al. A novel potent vasoconstrictor peptide produced by vascular endothelial cells. Nature 1988; 332: 411-415.
5. Inoue A, Yanagisawa M, Kimura S et al. The human endothelin family: three structurally and pharmacologically distinct isopeptides predicted by three separate genes. Proc Nat Acad Sci USA 1989; 86: 2863-2867.

6. Walder CE, Thomas GR, Thiemermann C et al. The hemodynamic effects of endothelin-1 in the pithed rat. J Cardiovasc Pharmacol 1989; 13 (Suppl 5): S93-S97.

7. Gardiner SM, Compton AM, Bennett T. Regional haemodynamic effects of endothelin-1 and ET-3 in conscious Long Evans and Brattleboro rats. Br J Pharmacol 1990; 99: 107-112.

8. Whittle BJR, Payne AN, Esplugues JV. Cardiopulmonary and gastric ulcerogenic actions of endothelin-1 in the guinea pig and rat. J Cardiovasc Pharmacol 1989; 13 (Suppl 5): S103-S107.

9. Braquet P, Touvay C, Lagente V et al. Effect of endothelin-1 on blood pressure and bronchopulmonary system of the guinea pig. J Cardiovasc Pharmacol 1989; 13 (Suppl 5): S143-S146.

10. Miyamori I, Itoh Y, Matsubara T et al. Systemic and regional effects of endothelin in rabbits: effects of endothelin antibody. Clin Exp Pharmacol Physiol 1990; 17(10): 691-696.

11. Clarke JG, Larkin SW, Benjamin N et al. Endothelin-1 is a potent long-lasting vasoconstrictor in dog peripheral vasculature in vivo. J Cardiovasc Pharmacol 1989; 13 (Suppl 5): S211-S212.

12. Pernow J, Franco-Cereceda A, Matran R et al. Effect of endothelin-1 on regional vascular resistances in the pig. J Cardiovasc Pharmacol 1989; 13 (Suppl 5): S205-S206.

13. Dieguez G, Garcia JL, Fernandez N et al. Cerebrovascular and coronary effects of endothelin-1 in the goat. Am J Physiol 1992; 263: R834-R839.

14. Olson KR, Duff DW, Farrell AP et al. Cardiovascular effects of endothelin in trout. Am J Physiol 1991; 260: H1214-H1223.

15. Lerman A, Hildebrand FL, Aarhus LL et al. Endothelin has biological actions at pathophysiological concentrations. Circulation 1991; 83: 1808-1814.

16. Mortensen LH, Pawloski CM, Kanagy NL et al. Chronic hypertension produced by infusion of endothelin in rats. Hypertension 1990; 15: 729-733.

17. Mortensen LH, Fink GD. Salt-dependency of endothelin-induced, chronic hypertension in conscious rats. Hypertension 1992; 19: 549-554.

18. Anggård E, Galton S, Rae G et al. The fate of radioiodinated endothelin-1 and endothelin-3 in the rat. J Cardiovasc Pharmacol 1989; 13 (Suppl 5): S46-S49.

19. Shiba R, Yanagisawa M, Miyauchi T et al. Elimination of intravenously injected endothelin-1 from the circulation of the rat. J Cardiovasc Pharmacol 1989; 13 (Suppl 5): S98-S101.

20. Hirata Y, Yoshimi H, Takaichi S et al. Binding and receptor down-regulation of a novel vasoconstrictor endothelin in cultured rat vascular smooth muscle cells. FEBS Lett 1988; 239: 13-17.

21. Kohno M, Murakawa K, Yasunari K et al. Prolonged blood pressure elevation after endothelin administration in bilaterally nephrectomised rats. Metabolism 1989; 38: 712-713.

22. Williams DL, Jones KL, Pettibone DJ et al. Sarafotoxin S6c: an agonist

which distinguishes between endothelin receptor subtypes. Biochem Biophys Res Commun 1991; 175: 556-561.

23. Douglas SA, Hiley CR. Endothelium-dependent mesenteric vasorelaxant effects and systemic actions of endothelin (16-21) and other endothelin-related peptides in the rat. Br J Pharmacol 1991; 104: 311-320.

24. Clozel M, Gray GA, Breu V et al. The endothelin ET_B receptor mediates both vasodilation and vasoconstriction in vivo. Biochem Biophys Res Commun 1992; 186: 867-873.

25. Gardiner SM, Kemp PA, March JE et al. Effects of an endothelin$_1$-receptor antagonist, FR139317, on regional haemodynamic responses to endothelin-1 and [Ala^{11,15}]Ac-endothelin-1 (6-21) in conscious rats. Br J Pharmacol 1994; 112: 477-486.

26. Moreland S, McMullen D, Abboa-Offei B et al. Evidence for a differential location of vasoconstrictor endothelin receptors in the vasculature. Br J Pharmacol 1994; 112: 704-708.

27. Ihara M, Fukoroda T, Saeki M et al. An endothelin receptor (ET_A) antagonist isolated from *Streptomyces misakiensis*. Biochem Biophys Res Commun 1991; 178: 132-137.

28. Bigaud M, Pelton JT. Discrimination between ET_A- and ET_B-receptor-mediated effects of endothelin-1 and [Ala ^{1,3,11,15}]endothelin-1 by BQ-123 in the anaesthetized rat. Br J Pharmacol 1992; 107: 912-918.

29. McMurdo L, Corder R, Thiemermann C et al. Incomplete inhibition of the pressor effects of endothelin-1 and related peptides in the anaesthetized rat with BQ-123 provides evidence for more than one vasoconstrictor receptor. Br J Pharmacol 1993; 108: 557-561.

30. Warner TD, Allcock GH, Vane JR. Reversal of established responses to endothelin-1 in vivo and in vitro by the endothelin receptor antagonists, BQ-123 and PD 145065. Br J Pharmacol 1994; 112: 207-213.

31. Gardiner SM, Kemp PA, March JE, Bennett T. Effects of bosentan (Ro 47-0203), an ET_A-, ET_B-receptor antagonist, on regional haemodynamic responses to endothelins in conscious rats. Br J Pharmacol 1994b; 112: 823-830.

32. Gardiner SM, Compton AM, Bennett T. Effect of indomethacin on the regional haemodynamic responses to low doses of endothelins and sarafotoxin. Br J Pharmacol 1990; 100: 158-162.

33. Horgan MJ, Pinheiro, JMB, Malik, AB. Mechanism of endothelin-1 induced pulmonary vasoconstriction. Circ Res 1991; 69: 157-164.

34. Toga H, Raj JU, Hillyard, R et al. Endothelin effects in isolated, perfused lamb lungs: role of cyclo-oxygenase inhibition and vasomotor tone. Am J Physiol 1991; 261: H443-H450.

35. Kurose I, Miura S, Suematsu M et al. Involvement of platelet-activating factor in endothelin-induced vascular smooth muscle cell contraction. J Cardiovasc Pharmacol 1991; 17 (Suppl 7): S279-S283.

36. Filep JG, Sirois MG, Rousseau A et al. Effects of endothelin-1 on vascular permeability in the conscious rat: interactions with platelet-activating factor. Br J Pharmacol 1991b; 104: 797-804.

37. Vierhapper H, Wagner O, Nowotny P et al. Effect of endothelin-1 in man. Circulation 1990; 81: 1415-1418.

38. Weitzberg E, Ahlborg G, Lundberg JM. Long-lasting vasoconstriction and efficient regional extraction of endothelin-1 in human splanchnic and renal tissues. Biochem Biophys Res Commun 1991; 180: 1298-1303.

39. Wagner OF, Vierhapper H, Gasic S et al. Regional effects and clearance of endothelin-1 across pulmonary and splanchnic circulation. Eur J Clin Invest 1992; 22: 277-282.

40. Vierhapper H, Wagner OF, Nowotny P et al. Effect of endothelin-1 in man: pretreatment with nifedipine, with indomethicin and with cyclosporine A. Eur J Clin Invest 1992; 22: 55-59.

41. Gasic S, Wagner OF, Vierhapper H et al. Regional haemodynamic effects and clearance of endothelin-1 in humans: renal and peripheral tissues may contribute to the overall disposal of the peptide. J Cardiovasc Pharmacol 1992; 19: 176-180.

42. Sørensen SS, Madsen JK, Pedersen EB. Systemic and renal effect of intravenous infusion of endothelin-1 in healthy human volunteers. Am J Physiol 1994; 266: F411-F418.

43. Kitazumi K, Shiba T, Nishiki K et al. Vasodilator effects of sarafotoxins and endothelin-1 in spontaneously hypertensive rats and rat isolated perfused mesentery. Biochem Pharmacol 1990; 40: 1843-1847.

44. Ohlstein EH, Vickery L, Sauermelch C et al. Vasodilation induced by endothelin: role of EDRF and prostanoids in rat hindquarters. Am J Physiol 1990; 259: H1835-H1841.

45. Mortensen LH, Fink GD. Hemodynamic effect of human and rat endothelin administration into conscious rats. Am J Physiol 1990; 258: H362-H368.

46. Gardiner SM, Kemp PA, Bennett T. Regional haemodynamic responses to intravenous and intraarterial endothelin-1 and big endothelin-1 in conscious rats. Br J Pharmacol 1993; 110: 1532-1536.

47. Whittle BJR, Lopez-Belmonte J, Rees DD. Modulation of the vasodepressor actions of acetylcholine, bradykinin, substance P and endothelin in the rat by a specific inhibitor of nitric oxide formation. Br J Pharmacol 1989; 98: 646-652.

48. Gardiner SM, Compton AM, Kemp PA, Bennett T. Regional and cardiac haemodynamic responses to glyceryl trinitrate, acetylcholine, bradykinin and endothelin-1 in conscious rats: effects of N^G-nitro-L-arginine methyl ester. Br J Pharmacol 1990; 101: 632-639.

49. Auberson S, Lacroix JS, Morel DR et al. Methylene blue attenuates vasodilation and enhances vasoconstriction in response to endothelin-1 in the pig nasal mucosa. Acta Physiol Scand 1991; 142: 149-156.

50. Fozard JR, Part ML. The role of nitric oxide in the regional vasodilator effects of endothelin-1 in the rat. Br J Pharmacol 1992; 105: 744-750.

51. Lerman A, Sandok EK, Hildebrand FL Jr et al. Inhibition of endothelium-derived relaxing factor enhances endothelin-mediated vasoconstriction. Circulation 1992; 85: 1894-1898.

52. Rogerson ME, Cairns HS, Fairbanks LD et al. Endothelin-1 in the rabbit: interactions with cyclo-oxygenase and NO-synthase products. Br J Pharmacol 1993; 108: 838-843.

53. Filep JG, Foldes-Filep E, Rousseau A et al. Vascular responses to endothelin-1 following inhibition of nitric oxide synthesis in the conscious rat. Br J Pharmacol 1993; 110: 1213-1221.

54. Granstam E, Wang L, Bill A. Vascular effects of endothelin-1 in the cat: modification by indomethacin and L-NAME. Acta Physiol Scand 1993; 148: 165-176.

55. Filep JG, Herman F, Battistini B et al. Antiaggregatory and hypotensive effects of endothelin-1 in beagle dogs: role for prostacyclin. J Cardiovasc Pharmacol 1991; 17 (Suppl 7): S216-S218.

56. De Nucci G, Thomas R, D'Orleans-Juste P et al. Pressor effects of circulating endothelin are limited by its removal in the pulmonary circulation and by release of prostacyclin and endothelium-derived relaxing factor. Proc Nat Acad Sci USA 1988; 85: 9797-9800.

57. Cirino M, Battistini B, Yano M et al. Dual cardiovascular effects of endothelin-1 dissociated by BQ-153, a novel ET_A receptor antagonist. J Cardiovasc Pharmacol 1994; 24: 587-594.

58. Gardiner SM, Compton AM, Bennett T. N^G-monomethyl-L-arginine does not inhibit the hindquarters vasodilator action of endothelin-1 in conscious rats. Eur J Pharmacol 1989; 171: 237-240.

59. Fozard JR, Part ML. No major role for atrial natriuretic peptide in the vasodilator response to endothelin-1 in the spontaneously hypertensive rat. Eur J Pharmacol 1990; 180: 153-159.

60. Kurihara H, Yoshizumi M, Sugiyama T et al. The possible role of endothelin-1 in the pathogenesis of coronary vasospasm. J Cardiovasc Pharmacol 1989; 13 (Suppl 5): S132-S137.

61. Miller WL, Redfield MM, Burnett JC Jr. Integrated cardiac, renal and endocrine actions of endothelin. J Clin Invest 1989; 83: 317-320.

62. Wang J, Zeballos GA, Kaley G et al. Dilation and constriction of large coronary arteries in conscious dogs by endothelin. Am J Physiol 1991; 261: H1379-H1386.

63. Hirata K, Matsuda Y, Akita H et al. Myocardial ischaemia induced by endothelin in the intact rabbit: angiographic analysis. Cardiovasc Res 1990; 24: 879-883.

64. Clozel M, Clozel JP. Effects of endothelin on regional blood flows in squirrel monkeys. J Pharmacol Exp Ther 1989; 250: 1125-1131.

65. Ezra D, Goldstein RE, Czaja JF et al. Lethal ischaemia due to intracoronary endothelin in pigs. Am J Physiol 1989; 257: H339-H343.

66. Yang X, Madeddu P, Micheletti R et al. Effects of intravenous endothelin on hemodynamics and cardiac contractility in conscious Milan normotensive rats. J Cardiovasc Pharmacol 1991; 17: 662-669.

67. Gardiner SM, Compton AM, Bennett T. Effects of endothelin-1 on cardiac output in conscious rats in the absence and presence of cardiac autonomic blockade. Eur J Pharmacol 1990; 183: 2232-2233.

68. Garcia R, Lachance D, Thibault G. Positive inotropic action, natriuresis and atrial natriuretic factor release induced by endothelin in the conscious rat. J Hypertension 1990; 8: 725-731.

69. Weiser E, Wollberg Z, Kochva E, Lee SY. Cardiotoxic effects of the venom of the burrowing asp, *Actractaspis engaddensis (Actractaspididae, Ophidia)*. Toxicon 1984; 22: 767-774.

70. Wright CE, Fozard JR. Regional vasodilation is a prominent feature of the haemodynamic response to endothelin in anaesthetized spontaneously hypertensive rats. Eur J Pharmacol 1988; 155: 201-203.

71. Han SP, Trapani AJ, Fok KF et al. Effects of endothelinon regional hemodynamics in conscious rats. Eur J Pharmacol 1989; 159: 303-305.

72. Knuepfer MM, Han SP, Trapani AJ et al. Regional hemodynamic and baroreflex effects of endothelin in rats. Am J Physiol 1989; 257: H918-H926.

73. Clarke JG, Benjamin N, Larkin SW et al. Endothelin is a potent long-lasting vasoconstrictor in men. Am J Physiol 1989; 257: H2033-H2035.

74. Hughes AD, Thom SAM, Woodall N et al. Human vascular responses to endothelin-1: observations in vivo and in vitro. J Cardiovasc Pharmacol 1989; 13 (suppl 5): S225-S228.

75. Dahlöf B, Gustaffsson D, Hedner T et al. Regional haemodynamic effects of endothelin-1 in rat and man: unexpected adverse reactions. J Hypertens 1990; 8: 811-817.

76. Pernow J, Hemsén A, Lundberg JM et al. Potent vasoconstrictor effects and clearance of endothelin in the human forearm. Acta Physiol Scand 1991; 141: 319-324.

77. Cockcroft JR, Clarke JG, Webb DJ. The effect of intra-arterial endothelin on resting blood flow and sympathetically mediated vasoconstriction in the forearm of man. Br J clin Pharmac 1991; 31: 521-524.

78. Kiowski W, Luscher TF, Linder L et al. Endothelin-1-induced vasoconstriction in humans: reversal by calcium channel blockade but not by nitrovasodilators or endothelium-derived relaxing factor. Circulation 1991; 83: 469-475.

79. Kiowski W, Linder L. Reversal of endothelin-1-induced vasoconstriction by nifedipine in human resistance vessels in vivo. Am J Cardiol 1992; 69: 1063-1066.

80. Andrawis NS, Gilligan J, Abernethy DR. Endothelin-1-mediated vasoconstriction: specific blockade by verapamil. Clin Pharmacol Ther 1992; 52: 583-589.

81. Haynes WG, Webb DJ. Contribution of endogenous generation of endothelin-1 to basal vascular tone. Lancet. 1994; 344: 852-854.

82. Haynes WG, Webb DJ. Venoconstriction to endothelin-1 in humans: the role of calcium and potassium channels. Am J Physiol 1993; 265: H1676-H1681.

83. Haynes WG, Strachan FE, Webb DJ. Endothelin ET_A and ET_B receptors cause vasoconstriction of human resistance and capacitance vessels in vivo. Circulation 1995; in press.

84. Wenzel RR, Noll G, Luscher TF. Endothelin receptor antagonists inhibit endothelin in human skin microcirculation. Hypertension 1994; 23: 581-586.

85. Haefeli WE, Linder L, Kiowski W et al. In vivo properties of endothelin-1 and endothelin-3 in human hand veins and its reversal by bradykinin and verapamil. Hypertension 1993; 22: 343 (abstract).

86. Haynes WG, Webb DJ. Endothelium-dependent modulation of responses to endothelin-1 in human veins. Clin Sci 1993; 84: 427-433.

87. Strachan FE, Haynes WG, Storey P, Webb DJ. Endothelin type A and B receptors mediate venoconstriction in humans. Br J Pharmacol 1995 (abstract in press)

88. King AJ, Brenner BM, Anderson S. Endothelin: a potent renal and systemic vasoconstrictor peptide. Am J Physiol 1989; 256: F1051-F1058.

89. Hoffman A, Grossman E, Keiser HR. Opposite effects of endothelin-1 and big-endothelin-(1-39) on renal function in rats. Eur J Pharmacol 1990; 182: 603-606.

90. Brooks DP, DePalma PD, Pullen M et al. Characterization of canine renal endothelin receptor subtypes and their function. J Pharmacol Exp Ther 1994; 268: 1091-1097.

91. Cristol J-P, Warner TD, Thiemermann C et al. Mediation via different receptors of the vasoconstrictor effects of endothelins and sarafotoxins in the systemic circulation and renal vasculature of the anaesthetized rat. Br J Pharmacol 1993; 108: 776-779.

92. Pollock DM, Opgenorth TJ. Evidence for endothelin-induced renal vasoconstriction independent of endothelin$_A$ receptor activation. Am J Physiol 1993; 264: R222-R226.

93. Cirino M, Motz C, Maw J et al. BQ-153, a novel ET (ET)$_A$ antagonist, attenuates the renal vascular effects of endothelin-1. J Pharm Pharmacol 1992; 44: 782-785.

94. Cao I, Banks RO. Cardiorenal actions of endothelin, part II: Effects of cyclo-oxygenase inhibitors. Life Sci 1990; 46: 585-590.

95. Munger KA, Takahashi K, Awazu M et al. Maintenance of endothelin-induced renal arteriolar constriction in rats is cyclo-oxygenase dependent. Am J Physiol 1993; 264: F637-644.

96. Perico N, Cornejo RP, Benigni A et al. Endothelin induces diuresis and natriuresis in the rat by acting on proximal tubular cells through a mechanism mediated by lipoxygenase products. J Am Soc Nephrol 1991; 2(1): 57-69.

97. Terada Y, Tomita K, Nonoguchi H et al. Different localisation of two types of endothelin receptor mRNA in microdissected rat nephron segments using reverse transcription and polymerase chain reaction assay. J Clin Invest 1992; 90: 107-112.

98. Rabelink TJ, Kaasjager, KAH, Boer, P et al. Effects of endothelin on renal function in humans: implications for physiology and pathophysiology. Kidney Int; 1994; 46: 376-381.

99. Lippton HL, Cohen GA, Knight M et al. Evidence for distinct endothelin receptors in the pulmonary vascular bed in vivo. J Cardiovasc Pharmacol 1991; 17 (Suppl 7): S370-S373.

100. Deleuze PH, Adnot S, Shiiya N et al. Endothelin dilates bovine pulmonary circulation and reverses hypoxic pulmonary vasoconstriction. J Cardiovasc Pharmacol 1992; 19: 354-360.

101. Wong J, Vanderford PA, Fineman JR et al. Endothelin-1 produces pulmonary vasodilation in the intact newborn lamb. Am J Physiol 1993; 265: H1318-H1325.

102. Lippton HL, Hauth TA, Summer WR et al. Endothelin produces pulmonary vasoconstriction and systemic vasodilation. J Appl Physiol 1989; 66: 1008-1012.

103. Kadowitz PJ, McMahon TJ, Hood JS et al. Pulmonary vascular and airway responses to endothelin-1 are mediated by different mechanisms in the cat. J Cardiovasc Pharmacol 1991; 17 (Suppl 7): S374-S377.

104. Filep JG. Endothelin peptides: biological actions and pathophysiological significance in the lung. Life Sci 1993; 52: 119-133.

105. Asano T, Ikegaki I, Satoh S et al. Endothelin: a potential modulator of cerebral vasospasm. Eur J Pharmacol 1990; 190: 365-372.

106. Salom JB, Torregrossa G, Miranda FJ et al. Effects of endothelin-1 on the cerebrovascular bed of the goat. Eur J Pharmacol 1991; 192: 39-45.

107. Ouchi Y, Kim S, Souza AC et al. Central effect of endothelin on blood pressure in conscious rats. Am J Physiol 1989; 256: H1747-H1751.

108. Yamamoto T, Kimura T, Ota K et al. Central effects of endothelin-1 on vasopressin and atrial natriuretic peptide release and cardiovascular and renal function in conscious rats. J Cardiovasc Pharmacol 1991; 17 (Suppl 7): S316-S318.

109. Matsumura K, Abe I, Tsuchihashi T et al. Central effect of endothelin on neurohormonal responses in conscious rabbits. Hypertension 1991; 17: 1192-1196.

110. Itoh S, van Den Busse H. Sensitization of baroreceptor reflex by central endothelin in conscious rats. Am J Physiol 1991; 260: H1106-H1112.

111. Nishimura M, Takahashi H, Matsusawa M et al. Chronic intracerebroventricular infusions of endothelin elevate arterial pressure in rats. J Hypertens 1991; 9: 71-76.

112. Samson WK, Skala KD, Alexander BD et al. Hypothalamic endothelin: presence and effects related to fluid and electrolyte homeostasis. J Cardiovasc Pharmacol 1991; 17 (Suppl 7): S346-S349.

113. Spyer KM, McQueen DS, Dashwood MR et al. Localization of [^{125}I]endothelin binding sites in the region of the carotid bifurcation and brainstem of the cat: possible baro- and chemoreceptor involvement. J Cardiovasc Pharmacol 1991; 17 (Suppl 7): S385-S389.

114. Wong-Dusting HK, La M, Rand MJ. Mechanisms of the effects of endothelin on responses to noradrenaline and sympathetic nerve stimulation. Clin Exp Pharmacol Physiol 1990; 17: 269-273.

115. Yang ZH, Richard V, von Segesser L et al. Threshold concentrations of endothelin-1 potentiate contractions to norepinephrine and serotonin in human arteries: a new mechanism of vasospasm? Circulation. 1990; 82: 188-195.

116. Waite RP, Pang CCY. The sympathetic nervous system facilitates endothelin-1 effects on venous tone. J Pharmacol Exp Ther 1992; 260: 45-50.

117. Haynes WG, Hand M, Johnstone H et al. Direct and sympathetically mediated venoconstriction in essential hypertension: enhanced responses to endothelin-1. J Clin Invest 1994; 94: 1359-1364.

118. Rakugi H, Nakamaru M, Saito H et al. Endothelin inhibits renin release from isolated rat glomeruli. Biochem Biophys Res Commun 1988; 155: 1244-1247.

119. Kawaguchi H, Sawa H, Yasuda H. Effect of endothelin on angiotensin converting enzyme in cultured pulmonary artery endothelial cells. J Hypertens 1991; 9: 171-174.

120. Rakugi H, Tabuchi Y, Nakamaru M et al. Endothelin activates the vascular renin-angiotensin system in rat mesenteric arteries. Biochem Int 1990; 21: 867-872.

121. Cozza EN, Gomez-Sanchez CE, Foecking M, Chiou S. Endothelin binding to cultured calf adrenal zona glomerulosa cells and stimulation of aldosterone secretion. J Clin Invest 1989; 84: 1032-1035.

122. Boarder MR, Marriott DB. Characterization of endothelin-1 stimulation of catecholamine release from adrenal chromaffin cells. J Cardiovasc Pharmacol 1989; 13 (Suppl 5): S223-S224.

123. Nakamoto H, Suzuki H, Murakami M et al. Effects of endothelin on systemic and renal hemodynamics and neuroendocrine hormones in conscious dogs. Clin Sci 1989; 77: 567-572.

124. Cao L, Banks RO. Cardiorenal actions of endothelin, part I: effects of converting enzyme inhibition. Life Sci 1990b; 46: 577-583.

125. Mortensen LH, Fink GD. Captopril prevents chronic hypertension produced by infusion of endothelin-1 in rats. Hypertension 1992; 19: 676-680.

126. Kaufmann H, Oribe E, Oliver JA. Plasma endothelin during upright tilt: relevance for orthostatic hypotension? Lancet 1991; 338: 1542-1545.

127. Stasch J-P, Hirth-Dietrich C, Kazda S et al. endothelin stimulates release of atrial natriuretic peptides in vitro and in vivo. Life Sci 1989; 45: 869-875.

128. Valentin JP, Gardner DG, Wiedemann E et al. Modulation of endothelin effects on blood pressure and hematocrit by atrial natriuretic peptide. Hypertension 1991; 17: 864-869.

129. Horio T, Kohno M, Takeda T. Effects of arginine vasopressin, angiotensin II and endothelin-1 on the release of brain natriuretic peptide in vivo and in vitro. Clin Exp Pharmacol Physiol 1992; 19: 575-582.

130. Kashiwabara T, Inagaki Y, Ohta H et al. Putative precursors of endothelin have less vasoconstrictor activity in vitro but a potent pressor effect in vivo. FEBS Lett 1989; 247: 73-76.

131. Gardiner SM, Compton AM, Kemp PA et al. The effects of phosphoramidon on the regional haemodynamic responses to human proendothelin-1 [1-38] in conscious rats. Br J Pharmacol 1991; 103: 2009-2015.

132. McMahon EG, Palomo MA, Moore WM et al. Phosphoramidon blocks the pressor activity of big endothelin-(1-39) in vivo and conversion of big endothelin-(1-39) to endothelin-(1-21) in vitro. Proc Natl Acad Sci USA 1991; 88: 703-707.

133. McMahon EG, Palomo MA, Moore WM. Phosphoramidon blocks the pressor activity of big endothelin[1-39] and lowers blood pressure in spontaneously hypertensive rats. J Cardiovasc Pharmacol 1991; 17 (Suppl 7): S29-S33.

134. Matsumura Y, Hisaki K, Takaoka M et al. Phosphoramidon, a metalloproteinase inhibitor, supresses the hypertensive effect of big endothelin-1. Eur J Pharmacol 1990; 185: 103-106.

135 Pollock DM, Opgenorth TJ. Evidence for metalloprotease involvement in the in vivo effects of big endothelin-1. Am J Physiol 1991; 261: R257-R263.

136. Fukuroda F, Noguchi K, Tsuchida S et al. Inhibition of biological actions of big-endothelin-1 by phosphoramidon. Biochem Biophys Res Commun 1990; 172: 390-395.

137. D'Orleans-Juste P, Telemaque S, Claing A. Different pharmacological profiles of big-endothelin-3 and big-endothelin-1 in vivo and in vitro. Br J Pharmacol 1991; 104: 440-444.

138. Hemsén A, Pernow J, Lundberg JM. Regional extraction of endothelins and conversion of big ET to ET-1 in the pig. Acta Physiol Scand 1991; 141: 325-334.

139. Gardiner SM, Kemp PA, Bennett T. Inhibition by phosphoramidon of the regional haemodynamic effects of proendothelin-2 and-3 in conscious rats. Br J Pharmacol 1992b; 107: 584-590.

140. Pollock DM, Divish BJ, Milicic I et al. In vivo characterization of a phosphoramidon-sensitive endothelin-converting enzyme in the rat. Eur J Pharmacol 1993; 231; 459-464.

141. Matsumura Y, Fujita K, Takaoka M et al. Big endothelin-3-induced hypertension and its inhibition by phosphoramidon in anaesthetized rats. Eur J Pharmacol 1993; 230:89-93.

142. Watanabe Y, Naruse M, Monzen C et al. Is big endothelin converted to endothelin-1 in circulating blood? J Cardiovasc Pharmacol 1991; 17 (Suppl 7): S503-S505.

143. Plumpton, C, Haynes, WG, Webb, DJ, Davenport, AP. Phosphoramidon inhibits the in vivo conversion of big ET-1 to ET-1 in man. Br J Pharmacol 1995; (abstract: in press)

144. Xu D, Emoto N, Giaid A et al. ECE-1: A membrane bound metallo-protease that catalyses the proteolytic activation of big endothelin-1. Cell 1994; 78: 473-485.

145. Haynes WG, Davenport AP, Webb DJ. Endothelin: progress in physiology and pharmacology. Trends Pharmacol Sci 1993; 14: 225-228.

146. Paul M, Rettig R, Talsness CE et al. Transgenic rats expressing the human endothelin-2 gene: a new model to study endothelin regulation in vivo. J Hypertens 1994; 12 (suppl 3): S72 (abstract).

147. Kurihara Y, Kurihara H, Suzuki H et al. Elevated blood pressure and craniofacial abnormalities in mice deficient in endothelin-1. Nature 1994; 368: 703-710.

148. Nishikibe M, Tsuchida S, Okada M et al. Antihypertensive effect of a newly synthesized endothelin antagonist, BQ-123, in a genetic hypertensive model. Life Sci 1993; 52:717-724.

149. Bazil MK, Lappe RW, Webb RL. Pharmacologic characterization of an endothelin$_A$ (ET$_A$) receptor antagonist in conscious rats. J Cardiovasc Pharmacol 1992; 20: 940-948.

150. Clozel M, Breu V, Burri K et al. Pathophysiological role of endothelin revealed by the first orally active ET receptor antagonist. Nature 1993; 365: 759-761.

151. Véniant M, Clozel JP, Hess P et al. Endothelin plays a role in the maintenance of blood pressure in normotensive guinea pigs. Life Sci 1994; 55: 445-454.

152. Erdos EG, Skidgell RA. Neutral endopeptidase 24.11 (enkephalinase) and related regulators of peptide hormones. FASEB J 1989; 3: 145-151.

153. Abassi ZA, Golomb E, Bridenhaugh R et al. Metabolism of endothelin-1 and big endothelin-1 by recombinant neutral endopeptidase EC.3.4.24.11. Br J Pharmacol 1993; 109: 1024-1028.

154. Bevan EG, Connell JMC, Doyle J et al. Candoxatril, a neutral endopeptidase inhibitor: efficacy and tolerability in essential hypertension. J Hypertens 1992; 10: 607-613.

155. Vallance P, Collier J, Moncada S. Effects of endothelium derived nitric oxide on peripheral arteriolar tone in man. Lancet 1989; ii: 997-1000.

CHAPTER 7

THE ENDOTHELINS IN CARDIOVASCULAR PATHOPHYSIOLOGY

David J. Webb, William G. Haynes

In early studies to examine the role of ETs in the pathophysiology of cardiovascular disease, before ECE inhibitors and ET receptor antagonists were available, much emphasis was placed on tissue and plasma immunoreactive ET concentrations. Although tissue concentrations may give valuable information, the interpretation of plasma concentrations remains difficult, depending not only on generation of ET, but also on receptor-mediated clearance and enzymatic degradation. ET-1 is predominantly released abluminally from the vascular endothelium[1] and it is unlikely that ET-1 acts as a circulating hormone, except where concentrations are extremely high.[2] In some studies, sensitivity to the ETs has been examined. However, because receptor number may be down-regulated by increased ET-1 concentrations,[3] unless responses to exogenous ET are interpreted in the light of ET receptor number and affinity, the results of such studies may suggest that ET is not involved in pathological conditions in which its production is increased. Another potential confounding factor, when examining responses in blood vessels, is the development of the structural change that can occur in hypertension and heart failure. These arguments are rehearsed within the section on hypertension but are also pertinent elsewhere. More recently, with the advent of the ECE inhibitor, phosphoramidon, and, more importantly, highly specific antagonists at ET receptors, it has been possible to define more clearly the role of the ET system in pathophysiology. This chapter addresses

Molecular Biology and Pharmacology of the Endothelins edited by Gillian A. Gray and David J. Webb. © 1995 R.G. Landes Company.

the pathological conditions in which the ET system has been implicated and in which inhibition of the production or effects of ET might offer new therapeutic perspectives, with particular emphasis on its role in the cardiovascular system.

HYPERTENSION

Established essential hypertension is a condition characterized by a high blood pressure in association with increased peripheral vascular resistance. Cardiac and vascular hypertrophy may also develop. In the early stages, and in borderline hypertension, cardiac output appears to increase[4,5] and sympathetic tone may be enhanced.[6] This common condition, affecting around 10% of the adult population, makes a major contribution to the population burden from cardiovascular disease, predisposing particularly to myocardial infarction, stroke, and heart and kidney failure. A number of effective hypotensive drugs are available for the treatment of this condition although none have much impact on the most common consequence of essential hypertension, myocardial infarction. In addition, the mechanisms causing essential hypertension are still poorly understood and the drugs presently available may not act to oppose the causal mechanisms in this polygenic disorder.

The endothelium is an attractive target for investigation in essential hypertension, and in cardiovascular disease in general. It is clear that nitric oxide plays an important role in regulation of vascular tone and platelet aggregation and several investigators have shown that endothelial dysfunction related to the L-arginine/nitric oxide system occurs in hypertension,[7-9] diabetes mellitus[10] and hyperlipidemia,[11] all of which conditions predispose to an increased risk of cardiovascular disease. ET-1 is also an attractive target for investigation in hypertension because it has potent vasoconstrictor and pressor properties, is mitogenic and can cause vessel hypertrophy, and it appears to enhance sympathetic function in vitro.

Endothelin Concentrations and Production

In animal models of hypertension, ET-1 concentrations are not raised unless accelerated hypertension is present, where they are positively correlated with plasma creatinine.[12-15] Local mesenteric vascular generation of ET appears to be increased in vitro in SHR but not normotensive WKY rats.[16] However, others have found ET immunoreactivity to be increased in blood vessels from deoxycorticosterone acetate (DOCA)-salt rats, but decreased in blood vessels from SHR as compared to normotensive WKY rats.[17] In functional studies, there was no difference in the hypotension following infusion of the ECE inhibitor, phosphoramidon, between WKY and SHR.[18] Also against a role for increased generation of ET-1 in the pathophysiology of experimental hypertension is the finding that polymorphisms of the preproET-1 gene do not co-segregate with blood pressure or cardiac weight in inbred Dahl rats.[19] Interestingly, the ET-3 gene does appear to be linked to a locus that regulates blood pressure in these rats.

Renal effects of ET may also potentially contribute to the development of hypertension. The vascular/glomerular actions (mainly ET_A) predispose to sodium retention[20] and may be enhanced in the SHR.[21] The renal tubular actions (mainly ET_B), act to increase urinary sodium and water excretion to salt and water excretion.[22-24] Generation of ET-1 in the renal medulla, particularly the collecting duct, is reduced in SHR compared to WKY rats[25,26] and may, therefore, contribute to sodium retention and hypertension in this species. Patients with essential hypertension excrete less immunoreactive ET in urine than normotensive controls,[27] emphasizing the need for further investigation in this area.

Plasma immunoreactive ET concentrations were reported to be elevated in early studies in hypertensive patients.[28-30] However, clearance of ET-1 is dependent on renal function,[31,32] and the very high concentrations found in severe and accelerated phase hypertension are, at least in part, secondary to impaired renal clearance. Studies in well-characterized hypertensive patients with normal renal function have shown similar concentrations of ET-1 to those in well-matched normotensive subjects.[33-36] Indeed, in one study there was a negative correlation between blood pressure and plasma ET-1 in the hypertensive group,[33] making a global increase in generation of ET-1 unlikely as a cause of essential hypertension. In the hypertension associated with pre-eclampsia of pregnancy, plasma ET is elevated despite normal renal function, consistent with a role for ET in the pathophysiology of this hypertensive condition.[37]

Increased production of ET-1 also occurs in one secondary form of hypertension, albeit rare. In 1991, Yokokawa and colleagues[38] described two cases of the skin tumor, hemangio-endothelioma, in which hypertension was associated with increased plasma ET-1 concentrations (Fig. 7.1). Biopsies of tumor cells displayed increased expression of mRNA for prepro ET-1, and strong immunohistochemical staining for the peptide. Blood pressure and plasma ET-1 concentrations returned to normal in both cases following surgical resection of the tumors and, in one patient, recurrence of the tumor led to further increase in both blood pressure and plasma ET-1.

SENSITIVITY TO ENDOTHELIN

In studies designed to examine vascular sensitivity to ET-1 in hypertension it is important to be aware that the development of vascular hypertrophy will tend to amplify responses in the hypertensive vessels.[39] In studies comparing WKY and SHR, both conduit and resistance vessels from the SHR are more sensitive to the effects of ET-1.[40-43] In the Dahl salt sensitive rat, vascular responsiveness to ET is enhanced

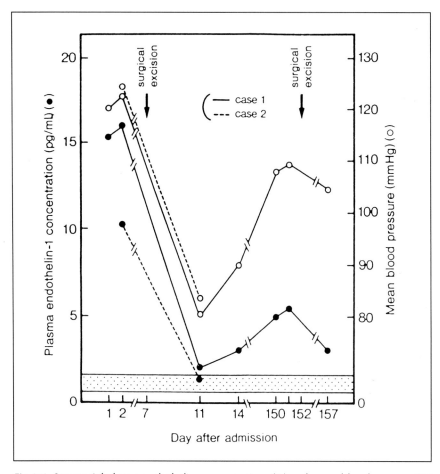

Fig. 7.1. Sequential plasma endothelin concentrations (●) and mean blood pressure (○) in two patients with malignant hemangio-endothelioma. The shaded area indicates the normal range of plasma endothelin-1 concentrations in humans. Adapted with permission from Yokokawa et al, Ann Intern Med 1991; 114: 213-215.

prior to, but not after, the development of hypertension.[44] Other investigators have reported decreased sensitivity to ET in the aorta and mesenteric resistance arteries from SHR,[45] DOCA-salt rats[46] and renovascular hypertensive animals.[47,48] It is possible that decreased sensitivity to ET may be related to down-regulation of specific ET receptors secondary either to increased local generation of ET or raised blood pressure.[17] Down-regulation of receptors is suggested by the decreased Ca^{2+} response to ET-1 of isolated vascular smooth muscle cells from SHR.[49] In this regard it is interesting that decreased sensitivity to ET in hypertensive animals can be restored to normal by antihypertensive therapy.[50]

Systemic doses of ET-1 in vivo have greater pressor effects in SHR than WKY rats[51] and in renovascular hypertensive than normotensive rabbits.[48] The mechanism for this enhanced sensitivity to ET-1 is unclear, because the number of binding sites for ET-1 in aortic smooth muscle[41] and heart[52] are lower in SHR, suggesting increased post-receptor sensitivity or the influence of vascular hypertrophy. There is, however, a relative increase in the number of binding sites for ET-1 in the brain of SHR, as compared to WKY rats,[52] so there may be increased central nervous system sensitivity to the peptide in hypertension.

In patients with essential hypertension, in vitro efficacy of ET in subcutaneous resistance arteries appears to be reduced.[53] However, in vivo, the situation is rather different. A recent study[36] has investigated whether vasoconstriction to ET-1 is altered in patients with essential hypertension as compared to well-matched normotensive control subjects. Responses were examined to local infusion of ET-1 into hand veins, rather than arteries, because the venous system may contribute to the high cardiac output reported to occur in early hypertension and because the confounding effects of vascular hypertrophy do not occur in hand veins.[36,54] Sympathetic responses were also examined in this study because ET-1 has been shown to potentiate sympathetic responses in vitro.[55-58] ET-1 caused a slow-onset venoconstriction in both groups of subjects, with the maximal effect reached by 90 minutes. However, maximal venoconstriction to ET-1 was substantially greater in hypertensive patients than in control subjects (Fig. 7.2). Sympathetically-mediated venoconstriction was also enhanced in the hypertensive patients, but not in the control subjects (Fig. 7.3).

These studies have shown that patients with essential hypertension have enhanced venoconstriction to ET-1.[36] These results might be explained by decreased local venous ET-1 generation although plasma ET concentrations were similar to those in normotensive subjects. The negative correlation with blood pressure in the normotensive subjects (Fig. 7.4) is against this phenomenon occurring solely as a consequence of the increase in blood pressure and a causal relationship with the elevation of blood pressure is supported by the positive correlation in the hypertensive subjects. Enhanced facilitation of sympathetic vasoconstriction appears to be a separate phenomenon because the lack of basal tone in these vessels precludes an explanation for enhanced venoconstriction to ET-1 by potentiated sympathetic responsiveness. In addition, the sympathetic effect appears to be confined to hypertensive patients because, as well as in this study, it was not seen in a separate study in normotensive subjects.[59]

STUDIES WITH INHIBITORS AND ANTAGONISTS

Recent studies support a role for ET-1 in experimental animal hypertension. Both ECE inhibitors and ET receptor antagonists lower blood pressure in hypertensive rat models.[18,60-63] The ET_A receptor antagonist BQ-123 acutely lowers blood pressure in stroke-prone SHR, but not in control SHR or WKY rats.[61] When administered chronically, BQ-123 prevents the development of stroke and renal abnormalities in stroke-prone SHR.[64] This ET_A receptor antagonist also lowers blood pressure in low renin animal models of hypertension (the SHR and DOCA/salt treated rats) but not in control animals or in a high renin animal model of hypertension.[60]

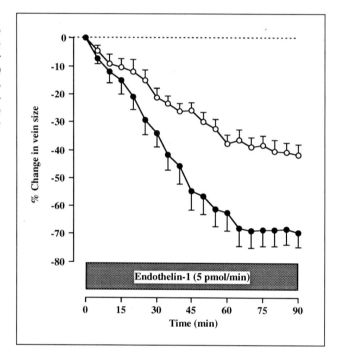

Fig. 7.2. Venoconstriction to endothelin-1 in normotensive (O) and hypertensive (●) subjects. Maximal venoconstriction to endothelin-1 was substantially greater in hypertensive patients (72 ± 5% at 90 minutes) than control subjects (41 ± 4% at 90 minutes; p = 0.001 v. hypertensives). Reproduced with permission from Haynes et al, J Clin Invest 1994; 94: 1359-1364.

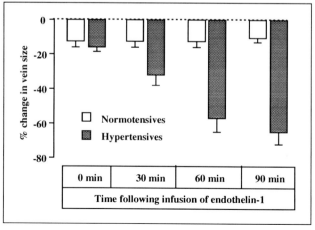

Fig. 7.3. Sympathetically mediated venoconstriction induced by single deep breath before and during infusion of endothelin-1 in hypertensive and normotensive subjects. Responses to deep breath were substantially enhanced in hypertensive (67 ± 7% at 90 minutes) but not normotensive subjects (11 ± 3% at 90 minutes) after infusion of endothelin-1 (p =0.001). This was observed whether venoconstriction was expressed as the absolute or percentage change in vein size. Reproduced with permission from Haynes et al, J Clin Invest 1994; 94: 1359-1364.

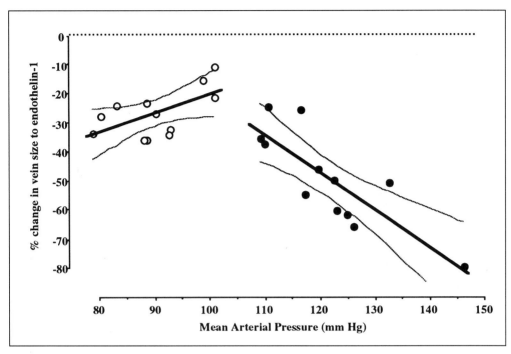

Fig. 7.4. Regression curves (with 95% confidence intervals) for the correlation between blood pressure and change in vein size to endothelin-1 in normotensive (○) and hypertensive (●) subjects. Reproduced with permission from Haynes et al, J Clin Invest 1994; 94: 1359-1364.

Although some studies have reported a reduction of blood pressure only in hypertensive animals it is important to remember that the absolute effects of all anti-hypertensive agents are proportionally greater the higher the pre-treatment blood pressure, and few studies have been of sufficient power to justify this interpretation. Certainly, although there is evidence that ET receptor antagonists cause vasodilatation in hypertensive patients,[65] preliminary evidence shows that ET receptor antagonists can also cause vasodilatation, and lower blood pressure in healthy normotensive humans.[66] At present, it seems unlikely that drugs based on blocking the ET system will be selectively 'anti-hypertensive' rather than hypotensive. However, based on vasodilator, anti-mitogenic and sympatholytic properties, they may offer particular advantages in the treatment of hypertension. Indeed, atherosclerotic human blood vessels show strong immunostaining for ET-1, and patients with atherosclerosis have plasma ET-1 concentrations that are raised in proportion to the number of affected vessels[67] so ET may contribute directly to the atherosclerotic process in vascular disease. A more recent report extends this observation by showing over-expression of ET-1 mRNA in the vascular smooth muscle of atherosclerotic human vessels.[68]

The development of ET receptor antagonists for the treatment of hypertension is likely to be a major focus for clinical investigation in the foreseeable future. As in other cardiovascular diseases, it remains to be seen whether an ET_A receptor antagonist, leaving endothelial ET_B receptors unopposed but vascular smooth muscle ET_B receptors unblocked, will prove superior to a combined antagonist at both receptors. ET_A receptor antagonists that also inhibit the vascular, but not endothelial, ET_B receptor would, in principle, seem ideal and, interestingly, there is already some preliminary evidence that these receptors can be distinguished pharmacologically (see chapter 4).

VASOSPASTIC DISORDERS

There are a number of vascular conditions which are characterized by instability of tone in small arteries, including Raynaud's disease, Prinzmetal's or variant angina and migraine, associated with vasospasm in digital, coronary and cerebral vessels, respectively. An association is recognized between variant angina, Raynaud's disease and migraine,[69] suggesting a shared abnormality of vascular function, possibly involving an imbalance between endothelium-derived dilator and constrictor factors. Indeed, circulating plasma ET concentrations are elevated in patients with Raynaud's phenomenon,[70,71] systemic sclerosis and coronary vasospasm,[72] between episodes of spasm, consistent with increased production. In Raynaud's phenomenon, circulating concentrations increase further following a cold challenge to a hand,[70] and ET-1 concentrations are higher still in venous blood draining the cold challenged hand, suggesting that it is involved in the vasospasm. Others have shown that specific ET binding sites, consistent with both ET_A and ET_B receptors, are present in the microvessels of human skin. These are increased in density in patients with systemic sclerosis, and not reduced in patients with Raynaud's phenomenon.[73] This group argues that failure of down-regulation of ET receptors in the face of increased circulating concentrations may contribute to the pathophysiology of Raynaud's. By contrast, in coronary vasospasm, ET-1 concentrations do not increase further during ischemic episodes,[72] suggesting that another factor, such as decreased nitric oxide production, may be responsible for initiating the attack. Although the evidence of a role for ET in these conditions remains relatively sparse, there may be grounds for further investigation in Raynaud's phenomenon, given that there are currently no particularly effective treatments and that this is a relatively common condition of cold climates.

Cerebral vasospasm, usually occurring between 4 and 10 days after the initial bleed, remains one of the major causes of morbidity and mortality in patients with aneurysmal sub-arachnoid hemorrhage. There is now a substantial body of evidence implicating ET in this condition. First, ET elicits a profound and sustained constriction of the cerebral vessels that undergo chronic constriction during subarachnoid haemorrhage.[74] Second, plasma and cerebrospinal fluid ET concentrations appear to be elevated in patients with subarachnoid hemorrhage, and are higher in those who later develop cerebral vasospasm,[75] although this finding has not been entirely reproducible.[76] Third, and very convincingly, in animal models of subarachnoid hemorrhage, ET receptor antagonists have been shown to both prevent (Fig. 7.5)[77] and reverse cerebral vasospasm.[78] Since reversal of vasospasm was produced in one case by an antagonist selective for the ET_A receptor,[78] it seems

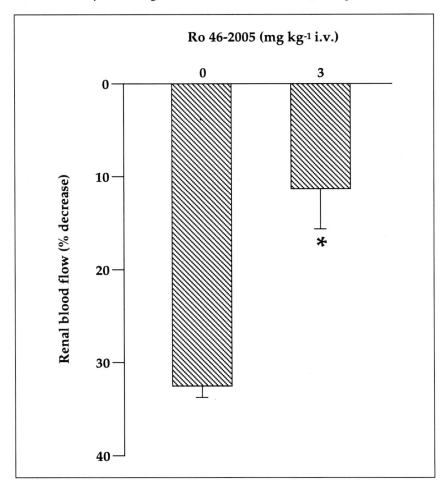

Fig. 7.5. Prevention by the combined $ET_{A/B}$ receptor antagonist, Ro 46-2005, of postischemic renal vasoconstriction in anesthetized and instrumented Wistar rats. Values shown are % decreases in renal blood flow expressed as mean ± SEM. *P<0.001 by Student's t-test. Renal ischemia was induced by a 45 minute clamping of the left renal artery by a snare placed around the artery at its origin. Ten minutes before induction of renal ischemia, the rats received an intravenous bolus injection of Ro 46-2005 (3 mg/kg; n=6 rats) or its saline vehicle (n=7 rats). On reperfusion, renal blood flow stabilized within 20-30 minutes. Adapted with permission from Clozel et al, Nature 1993; 365: 759-761.

likely that this receptor should be the major target in this condition. Given the findings in animal models, and the supportive data from patients, clinical studies must be awaited with great interest.

ISCHEMIC HEART DISEASE

Cardiac tissue immunoreactive ET-1 increases substantially (6 fold) after experimental myocardial infarction in rats,[79] with this tissue increase lasting longer than that in plasma.[80] In addition, the number of cardiac binding sites for ET-1 increase markedly following either ischemia alone, or ischemia with reperfusion.[81] Furthermore, pretreatment with anti-ET gamma globulin in a rat model of myocardial infarction causes a 40% decrease in infarct size.[79,80] Thus a combination of increased cardiac ET-1 content and additional binding sites for the peptide may contribute to extension of infarct size by compromising the blood flow in areas of reperfusion. However, reduction of myocardial infarction size has not been confirmed in more recent animal studies using selective ET_A or combined $ET_{A/B}$ receptor antagonists.[82,83] Other studies, of electrophysiological effects on the heart, suggest that ET-1 may act to oppose the arrhythmogenic effects of catecholamines and reduce the risk of dangerous ventricular arrythmias associated with myocardial infarction.[84]

In humans, plasma ET concentrations are normal in patients with stable angina and mild to moderate unstable angina,[85-87] although they are increased in patients with severe unstable angina, in whom they are associated with an increased risk of an adverse outcome.[88] Circulating concentrations of ET-1 and big ET-1 rise rapidly following myocardial infarction,[86] with concentrations several-fold higher than normal on the day of infarction; these return only slowly to normal over the next 7 to 10 days.[89] The duration and degree of elevation is proportional to the severity of myocardial infarction, as judged by Killip class, with the highest concentrations found in patients with cardiogenic shock.[85] These findings are consistent with the elevated ET-1 concentrations described in other forms of shock[90] and renal dysfunction, causing reduced clearance of ET, may contribute to this effect. Interestingly, high plasma ET concentrations three days after myocardial infarction are strongly and independently predictive of death in next 12 months.[91]

A number of interesting, but not wholly consistent, observations have been made in experimental myocardial infarction. Although plasma ET rises and is of prognostic significance, there is presently little evidence to justify clinical investigation of anti-ET drugs in patients with this condition. Given the wide range of drugs presently available for the treatment of myocardial infarction, and the likely diminishing returns with additional therapy, a more substantial body of evidence in favor of ET receptor antagonists will almost certainly be needed before such drugs will be developed clinically in this area.

CHRONIC HEART FAILURE

Chronic heart failure is a common condition with a high morbidity and mortality.[92] Increased peripheral and renal vascular resistance, associated with abnormal structure and tone in the peripheral blood vessels, is a feature of this condition. Of therecent therapeutic approaches to heart failure, vasodilatation has been the most effective.[93,94] Hence, the vasoconstrictor, mitogenic and anti-natriuretic properties of ET-1 have attracted attention to its potential pathophysiological role in chronic heart failure.

In animal models of heart failure, ET-1 concentrations rise three fold, and correlate closely with pulmonary capillary wedge pressure.[95] Interestingly, in the coronary artery ligation model of chronic heart failure in the rat, the combined $ET_{A/B}$ receptor antagonist, bosentan, not only reduced blood pressure, showing that ET plays a role in regulating blood pressure in this condition, but also had effects that were additive to those of ACE inhibition (Fig. 7.6).[96]

Plasma ET concentrations are increased two- to three fold in patients with chronic heart failure[97,98] and correlate closely with the degree of hemodynamic and functional impairment.[99-101] Big ET-1 in par-

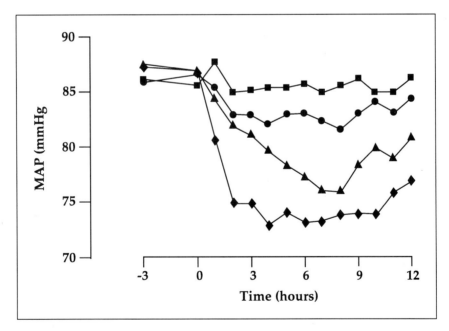

Fig. 7.6. Effects of placebo (■), the combined $ET_{A/B}$ receptor antagonist, bosentan (●), the ACE inhibitor, cilazapril (▲), and the combination of bosentan and cilazapril (◆) on mean arterial pressure (MAP) in rats with chronic heart failure. The combination of bosentan and cilazapril caused a further reduction in MAP compared with either bosentan (P<0.01) or cilazapril (P<0.05) alone. Values are the hourly means. Reproduced with permission from Teerlink et al, Circulation 1994; 90: 2510-2518.

ticular may be increased, suggesting increased ET synthesis rather than reduced clearance, in chronic heart failure.[100,101] Plasma ET concentration also relates to prognosis, with a higher concentration predicting a greater likelihood of death or need for cardiac transplantation.[100]

More recently, the ET_A receptor antagonist, BQ-123, and the ECE and NEP inhibitor, phosphoramidon, have been shown to cause substantial peripheral vasodilatation when infused into the forearm circulations of patients with stable chronic heart failure already receiving a loop diuretic and a maximal dose of an ACE inhibitor.[102] In contrast, the NEP inhibitor, thiorphan, caused vasoconstriction, indicating that ECE inhibition accounted for the vasodilatation caused by phosphoramidon. The role of ET_B receptors was not specifically addressed by this study but may be important. Circulating concentrations of angiotensin II are increased in heart failure, even after chronic treatment with an ACE inhibitor, and angiotensin II has been shown to down-regulate ET-1 binding sites[103] and up-regulate ET_B receptor mRNA.[104] Given that ET_B receptors can mediate vasoconstriction,[105] particularly in the venous system,[106,107] it may be that ET_B receptors are of greater functional significance with development of heart failure. Further studies with selective ET_B receptor antagonists and nonselective receptor antagonists are required to clarify this issue.

The recent observations by Love and colleagues,[102] that anti-ET therapy produces a significant additional reduction in vascular resistance in patients treated with full conventional therapy, indicate a need for preliminary acute and chronic hemodynamic studies in patients with heart failure. Even with optimal current treatment, the quality of life and prognosis of these patients remains poor.[98] Anti-ET therapy constitutes a very promising new development in the treatment of this condition that merits further clinical investigation.[98,102]

RENAL FAILURE

Acute renal failure due to post-ischemic acute tubular necrosis is characterized by intense renal vasoconstriction. This may be mediated by ET-1, which is a potent renal vasoconstrictor,[108] because production of, and sensitivity to, the peptide is increased by hypoxia.[81,109] A role for ET-1 is supported by the finding that, in animals, renal immunoreactive ET increases for at least 24 hours after 45 minutes of bilateral renal artery occlusion.[110] Also, renal vasoconstriction following ischemia is substantially ameliorated by administration of antibodies to ET, given either before[110,111] or after the insult.[112] Similar findings have been reported for BQ-123, a selective ET_A receptor antagonist[113,114] and Ro 46-2005, a combined ET_A/ET_B receptor antagonist (Fig. 7.7).[77] Circulating plasma concentrations of ET are increased in patients with post-ischemic renal failure.[115,116]

Acute renal failure commonly occurs in the presence of gram-negative septic shock. This condition is associated, overall, with a reduction in

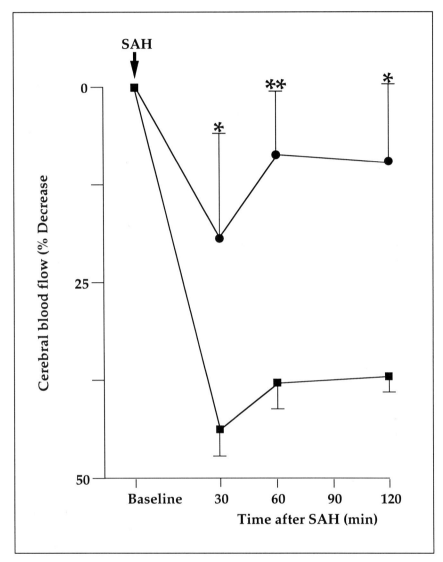

Fig. 7.7. Prevention by the combined $ET_{A/B}$ receptor antagonist, Ro 46-2005, of the decrease in cerebral blood flow (CBF) after induction of experimental subarachnoid hemorrhage (SAH) in rats. Ro 46-2005 (● : 3 mg/kg; n = 9 rats) or glucose as placebo (■ : n = 10 rats) was injected intravenously 10 minutes before SAH. The study was randomized and blinded. CBF was measured before and 30, 60 and 120 minutes after SAH using microspheres. Values shown are % changes in blood flow measured in the cerebellum expressed as mean ± SEM. *P < 0.05, **P < 0.01 by repeated measures analysis of variance. Adapted with permission from Clozel et al, Nature 1993; 365: 759-761.

systemic vascular resistance and arterial pressure, which is thought to be a consequence of enhanced nitric oxide production.[117] Paradoxically, in the face of widespread vasodilatation, renal vasoconstriction occurs, leading to acute renal failure.[118] Because endotoxin stimulates ET-1 release, both in vitro and in vivo,[119] and local administration of anti-ET gamma globulin completely reverses endotoxin induced renal vasoconstriction,[120] ET may contribute to this form of acute renal failure. Indeed, plasma ET concentrations are strikingly high in patients with septic shock.[90,121]

In animal models of chronic renal failure, there are increases in renal prepro ET-1 mRNA, cortical tissue immunoreactive ET-1 and urinary ET excretion.[122-124] In addition, renal ET generation is positively correlated with urinary protein excretion and glomerulosclerosis in such animals, supporting a role for ET in progression of chronic renal failure.[122,124] Furthermore, glomerular expression of mRNA for ET-1, and of ET_A and ET_B receptors is increased in experimental glomerulosclerosis.[125] Finally, administration of an ET receptor antagonist prevents the development of hypertension, glomerular damage and renal insufficiency in rats,[126] suggesting that ET contributes to the progression of chronic renal failure.

Plasma concentrations of immunoreactive big ET-1, ET-1, and ET-3 are also increased in chronic renal failure, and are raised two to four fold in patients on hemodialysis.[127,128] Although these increased concentrations may reflect impaired clearance of ET, the data from animal models, and the fact that urinary excretion of ET is also increased in patients with chronic renal failure,[129] suggest enhanced renal generation of the peptide in this disease. Plasma concentrations of ET may be sufficiently high to cause hypertension in chronic renal failure[2] and, in contrast to the situation in essential hypertension, ET-1 concentrations in hemodialysis patients are positively correlated with blood pressure.[130,131] Interestingly, plasma concentrations of ET-3, but not those of ET-1, rise further during hemodialysis;[130] this could reflect production of ET-3 by the central nervous system in response to dialysis-induced hypotension.

ET-1 may also contribute to the hypertension and renal impairment caused by cyclosporin. Production of ET-1 in vitro[132] and in vivo[133] is stimulated by cyclosporin, which also increases renal ET-1 receptor number.[134] Also, cyclosporin induced renal vasoconstriction is substantially attenuated by anti-ET gamma-globulin.[133] In addition, mesangial cell and renal afferent arteriolar contractile responses to cyclosporin are substantially attenuated by the ET_A receptor antagonist, BQ-123 in vitro[135,136] and in vivo.[137] The novel immunosuppressant, FK 506 (tacromilus), also appears to increase ET secretion in vitro.[138]

The evidence to date provides grounds for believing that anti-ET therapy might potentially be useful in both acute and chronic renal failure. However, first it will be essential to characterize the acute and chronic hemodynamic and renal effects of these drugs in healthy subjects and patients with stable renal failure. Currently there is no effective treatment for acute renal failure, a condition which carries a substantial mortality. Therefore, this might provide an important market for these agents, including the peptidic drugs requiring parenteral administration. Initial studies might concentrate on prophylaxis, either in situations recognized to cause acute renal failure, such as cytotoxic therapy or use of radiocontrast media, or in selected surgical patients at substantial risk of developing renal failure. Anti-ET therapy might also be useful in septic shock, but given that it may cause systemic vasodilatation this could only be realistic in the presence of a nitric oxide synthase inhibitor, possibly raising insuperable problems in clinical development. Investigation of these drugs in chronic renal failure would also be of interest but, as with the ACE inhibitor drugs, is likely only to occur after registration of a compound for one of the other, broader, indications.

FUTURE DIRECTIONS

There is now substantial experimental evidence linking ET with the pathophysiology of a range of cardiovascular diseases as diverse as essential hypertension, renal failure, chronic heart failure and sub-arachnoid hemorrhage. There is also supportive clinical evidence in all of these conditions, and it is also now clearly emerging that ET plays a fundamental pathophysiological role in regulation of vascular tone, indicating that clinical development programs concentrating on conditions associated with elevated vascular tone, such as hypertension and heart failure, may well be rewarded.

There are now a number of endothelin receptor antagonists—peptide and nonpeptide, selective for the ET_A receptor and nonselective, acting on both ET_A and ET_B receptors—coming into the early phases of clinical development. Although it appears likely that actions on both ET_A and ET_B receptors are necessary to completely block vasoconstriction to ET-1, given the presence of the ET_B receptor on the endothelial cells, it is still unclear which type of drug will emerge as the most effective, and this may well prove dependent on the pathology under consideration. In addition, with the cloning of ECE-1 it is likely, in time, that selective ECE inhibitors will emerge. These may also prove to have properties of benefit in the treatment of cardiovascular disease, and it may be that combined activity, for instance against ECE and ACE, may be more valuable than either action alone.

Drugs that block the ET system are the first major novel therapeutic target to emerge since the discovery of atriopeptidase inhibitors, and before those, the ACE inhibitors. It is difficult to be sure that ET antagonists will fulfill their promise, although the indications to date are extremely favorable. These are exciting times for both cardiovascular pharmacologists and cardiovascular physicians.

References

1. Yoshimoto S, Ishizaki Y, Sasaki T, Murota SI. Effect of carbon dioxide and oxygen on endothelin production by cultured porcine cerebral endothelial cells. Stroke 1991; 22: 378-383.

2. Webb DJ, Cockcroft JR. Plasma immunoreactive endothelin in uraemia. Lancet 1989; ii: 1211.

3. Hirata Y, Yoshimi H, Takachi S et al. Binding and receptor down-regulation of novel vasoconstrictor endothelin in cultured rat vascular smooth muscle cells. FEBS Lett 1988; 239: 13.

4. Ellis C, Julius S. Role of central blood volume in hyperkinetic borderline hypertension. Br Heart J 1973; 35: 450-455.

5. Safar M, London G, Levenson J et al. Rapid dextran infusion in essential hypertension. Hypertension 1979; 1: 615-623.

6. Anderson EA, Sinkey CA, Lawton WJ, Mark AL. Elevated sympathetic nerve activity in borderline hypertensive humans: direct evidence from intraneural recordings. Hypertension 1989; 14: 177-183.

7. Panza JA, Quyyumi AA, Brush JE, Epstein SE. Abnormal endothelium-dependent relaxation in patients with essential hypertension. N Engl J Med 1990; 323: 22-27.

8. Linder L, Kiowski W, Bühler FR, Lüscher TF. Indirect evidence for release of endothelium-dependent relaxing factor in human forearm circulation in vivo: blunted response in essential hypertension. Circulation 1990; 81: 1762-1767.

9. Calver A, Collier J, Moncada S, Vallance P. Effect of local intra-arterial N^G-monomethyl-L-arginine in patients with hypertension: the nitric oxide dilator mechanism appears abnormal. J Hypertens 1992; 10: 1025-1031.

10. Calver A, Collier J, Vallance P. Inhibition and stimulation of nitric oxide synthesis in the human forearm arterial bed of patients with insulin-dependent diabetes. J Clin Invest; 90: 2548-2554.

11. Chowienczyk PJ, Watts GF, Cockcroft JR, Ritter JM. Impaired endothelium-dependent vasodilation of forearm resistance vessels in hypercholesterolaemia. Lancet 1992; 340: 1430-1432.

12. Yan LS, Hui C, Yan GQ et al. Role of endothelin in the pathogenesis of hypertension in spontaneously hypertensive rats and 2 kidneys 1 clip rats. Chin Med J 1990; 103: 748-753.

13. Suzuki N, Miyauchi T, Tomobe Y et al. Plasma concentrations of endothelin in spontaneously hypertensive and DOCA-salt hypertensive rats. Biochem Biophys Res Commun 1990; 167: 941-947.

14. Vemulapalli S, Chui PJS, Rivelli M et al. Modulation of circulating endothelin levels in hypertension and endotoxaemia in rats. J Cardiovasc Pharmacol 1991; 18: 895-903.

15. Kohno M, Murakawa K-I, Horio T et al. Plasma immunoreactive endothelin-1 in experimental malignant hypertension. Hypertension 1991; 18: 93-100.

16. Miyamori I, Takeda Y, Yoneda T, Takeda R. Endothelin release from mesenteric arteries of spontaneously hypertensive rats. J Cardiovasc Pharmacol 1991; 17 (suppl 7): S408-S410.

17. Larivière R, Thibault G, Schiffrin EL. Increased endothelin-1 content in blood vessels of deoxycorticosterone acetate-salt hypertensive rats but not in spontaneously hypertensive rats. Hypertension 1993; 21: 294-300.

18. McMahon EG, Palomo MA, Moore WM. Phosphoramidon blocks the pressor activity of big endothelin[1-39] and lowers blood pressure in spontaneously hypertensive rats. J Cardiovasc Pharmacol 1991; 17 (Suppl 7): S29-S33.

19. Cicila GT, Rapp JP, Bloch KD et al. Cosegregation of the endothelin-3 locus with blood pressure and relative heart weight in inbred Dahl rats. J Hypertens 1994; 12: 643-651.

20. López-Farré A, Montanes I, Millas I, López-Nova JM. Effect of endothelin on renal function in rats. Eur J Pharmacol 1989; 163: 187-189.

21. Tomobe Y, Miyauchi T, Saito A et al. Effects of endothelin on the renal artery from spontaneously hypertensive and Wistar Kyoto rats. Eur J Pharmacol 1988; 152: 373-374.

22. Oishi R, Monoguchi H, Tomita K, Murumo F. Endothelin-1 inhibits AVP stimulated osmotic water permeability in rat inner medullary collecting duct. Am J Physiol 1991; 261: F951-F956.

23. Terada Y, Tomita K, Nonoguchi H, Marumo F. Different localization of two types of endothelin receptor mRNA in microdissected rat nephron segments using reverse transcription and polymerase chain reaction assay. J Clin Invest 1992; 90: 107-112.

24. Kohan DE, Padilla E. Osmolar regulation of endothelin-1 production by rat inner medullary collecting duct. J Clin Invest 1993; 91: 1235-1240.

25. Kitamura K, Tanaka T, Kato J et al. Immunoreactive endothelin in rat kidney inner medulla: marked decrease in spontaneously hypertensive rats. Biochem Biophys Res Commun 1989; 162: 38-44.

26. Hughes AK, Cline RC, Kohan DE. Alterations in renal endothelin-1 production in the spontaneously hypertensive rat. Hypertension 1992; 20: 666-673.

27. Hoffman A, Grossman E, Keiser HR. Opposite effects of endothelin-1 and big-endothelin-(1-39) on renal function in rats. Eur J Pharmacol 1990; 182: 603-606.

28. Kohno M, Yasunari K, Murakawa KI et al. Plasma immunoreactive endothelin in essential hypertension. Am J Med 1990; 88: 614-618.

29. Saito Y, Nakao K, Mukoyama M, Imura H. Increased plasma endothelin level in patients with essential hypertension. N Engl J Med 1990; 322: 205.

30. Shichiri M, Hirata Y, Ando K et al. Plasma endothelin levels in hypertension and chronic renal failure. Hypertension 1990; 15: 493-496.

31. Kohno M, Murakawa K, Yasunari K et al. Prolonged blood pressure elevation after endothelin administration in bilaterally nephrectomised rats. Metabolism 1989; 38: 712-713.

32. Koyama H, Tabata T, Nishzawa Y et al. Plasma endothelin levels in patients with uraemia. Lancet 1989; 333: 991-992.

33. Davenport AP, Ashby MJ, Easton P et al. A sensitive radioimmunoassay measuring endothelin-like immunoreactivity in human plasma: comparison of levels in patients with essential hypertension and normotensive control subjects. Clin. Sci. 1990; 78: 261-264.

34. Schiffrin EL, Thibault G. Plasma endothelin in human essential hypertension. Am J Hypertens 1991; 4: 303-308.

35. Haak T, Junmann E, Felber A et al. Increased plasma levels of endothelin in diabetic patients with hypertension. Am J Hypertens 1992; 5: 161-166.

36. Haynes WG, Hand MF, Johnstone HA et al. Direct and sympathetically mediated venoconstriction in essential hypertension: enhanced responses to endothelin. J Clin Invest 1994; 94: 1359-1364.

37. Florijn KW, Derkx FHM, Visser W et al. Plasma immunoreactive endothelin-1 in pregnant women with and without pre-eclampsia. J Cardiovasc Pharmacol 1991; 17 (suppl 7): S446-S448.

38. Yokokawa K, Tahara H, Kohno M et al. Hypertension associated with endothelin-secreting malignant haemangioendothelioma. Ann Intern Med 1991; 114: 213-215.

39. Folkow B. Cardiovascular structural adaption: its role in the initiation and maintenance of primary hypertension. Clin Sci Mol Med 1978; 55: 3S-22S.

40. Tomobe Y, Miyauchi T, Saito A et al. Effects of endothelin on the renal artery from spontaneously hypertensive and Wistar Kyoto rats. Eur J Pharmacol 1988; 152: 373-374.

41. Clozel M. Endothelin sensitivity and receptor binding in the aorta of spontaneously hypertensive rats. J Hypertens 1989; 7: 913-917.

42. MacLean MR, McGrath JC. Effects of endothelin-1 on isolated vascular beds from normotensive and spontaneously hypertensive rats. Eur J Pharmacol 1990; 190: 263-267.

43. Criscione L, Nellis P, Riniker B et al. Reactivity and sensitivity of mesenteric vascular beds and aortic rings of spontaneously hypertensive rats to endothelin: effects of calcium entry blockers. Br J Pharmacol 1990; 100: 31-36.

44. Goligorsky MS, Iijima K, Morgan M et al. Role of endothelin in the development of Dahl hypertension. J Cardiovasc Pharmacol 1991; 17 (suppl 7): S484-S491.

45. Dohi Y, Lüscher TF. Endothelin-1 in hypertensive resistance arteries: intraluminal and extraluminal dysfunction. Hypertension 1991; 18: 543-549.

46. Deng LY, Schiffrin EL. Effects of endothelin on resistance arteries of DOCA-salt hypertensive rats. Am J Physiol 1992; 262: H1782-H1787.

47. Dohi Y, Criscione L, Lüscher TF. Renovascular hypertension impairs formation of endothelium-derived relaxing factors and sensitivity to endothelin-1 in resistance arteries. Br J Pharmacol 1991; 104: 349-354.

48. Roberts-Thomson P, McRitchie RJ, Chalmers JP. Experimental hyperten-

sion produces diverse changes in the regional vascular responses to endothelin-1 in the rabbit and the rat. J Hypertens 1994; 12: 1225-1234.

49. Touyz RM, Tolloczko B, Schiffrin EL. Mesenteric vascular smooth muscle cells from spontaneously hypertensive rats display increased calcium reponses to angiotensin II but not to endothelin-1. J Hypertens 1994; 12: 663-673.

50. Dohi Y, Criscione L, Pfeiffer K, Lüscher TF. Normalization of endothelial dysfunction of hypertensive mesenteric resistance arteries by chronic therapy with benazapril or nifedipine. J Am Coll Cardiol 1992; 19 (suppl S): 226A.

51. Miyauchi T, Ishikawa T, Tomobe Y et al. (1989a). Characteristics of pressor response to endothelin in spontaneously hypertensive and Wistar-Kyoto rats. Hypertension 14: 427-434.

52. Gu XH, Casley DJ, Cincotta M, Nayler WG. ^{125}I-endothelin-1 binding to brain and cardiac membranes from normotensive and spontaneously hypertensive rats. Eur J Pharmacol 1990; 177: 205-209.

53. Schiffrin EL, Deng LY, Larochelle P. Blunted effects of endothelin upon small subcutaneous resistance arteries of mild essential hypertensive patients. J Hypertens 1992; 10: 437-444.

54. Eichler H-G, Ford GA, Blaschke TF et al. Responsiveness of superficial hand veins to phenylephrine in essential hypertension: alpha adrenergic blockade during prazosin therapy. J Clin Invest 1989; 83: 108-112.

55. Ouchi Y, Kim S, Souza AC et al. Central effect of endothelin on blood pressure in conscious rats. Am J Physiol 1989; 256: H1747-H1751.

56. Matsumura K, Abe I, Tsuchihashi T et al. Central effect of endothelin on neurohormonal responses in concious rabbits. Hypertension 1991; 17: 1192-1196.

57. Wong-Dusting HK, La M, Rand MJ. Mechanisms of the effects of endothelin on responses to noradrenaline and sympathetic nerve stimulation. Clin Exp Pharm Physiol 1990; 17: 269-273.

58. Yang ZH, Richard V, von Segesser L et al. Threshold concentrations of endothelin-1 potentiate contractions to norepinephrine and serotonin in human arteries: a new mechanism of vasospasm? Circulation 1990; 82: 188-195.

59. Cockcroft JR, Clarke JG, Webb DJ. The effect of intra-arterial endothelin on resting blood flow and sympathetically mediated vasoconstriction in the forearm of man. Br J Clin Pharmacol 1991; 31: 521-524.

60. Bazil MK, Lappe RW, Webb RL. Pharmacologic characterization of an endothelin$_A$ (ET$_A$) receptor antagonist in conscious rats. J Cardiovasc Pharmacol 1992; 20: 940-948.

61. Nishikibe M, Tsuchida S, Okada M et al. Antihypertensive effect of a newly synthetised endothelin antagonist, BQ123, in a genetic hypertensive model. Life Sci 1993; 52: 717-724.

62. McMahon EG, Palomo MA, Brown MA et al. Effect of phosphoramidon (endothelin converting enzyme inhibitor) and BQ-123 (endothelin receptor subtype A antagonist) on blood pressure in hypertensive rats. Am J Hypertens 1993; 6: 667-673.

63. Douglas SA, Gellai M, Ezekial M, Ohlstein EH. BQ-123, a selective endothelin subtype A-receptor antagonist, lowers blood pressure in different rat models of hypertension. J Hypertens 1994; 12: 561-567.

64. Nishikibe M, Okada M, Tsuchida S et al. Antihypertensive effect of a newly synthesised endothelin antagonist, BQ-123, in genetic hypertension models. J Hypertens 1992; 10 (suppl 4): S50.

65. Haynes WG, Strachan FE, Webb DJ. A physiological role for endothelin in maintenance of vascular tone in man. J Hypertens 1995; 12: (in press).

66. Haynes WG, Skwarski KM, Wyld PJ et al. Vasodilator effects of the $ET_{A/B}$ antagonist, TAK-044, in man. J Cardiovasc Pharmacol 1995 (in press).

67. Lerman A, Edwards BS, Hallett JW et al. Circulating and tissue endothelin immunoreactivity in advanced atherosclerosis. N Engl J Med 1991; 325: 997-1001.

68. Winkles JA, Alberts GF, Brogi E, Libby P. Endothelin-1 binding and endothelin receptor mRNA expression in normal and atherosclerotic human arteries. Biochem Biophys Res Commun 1993; 191: 1081-1088.

69. Miller D, Waters DD, Warnica W et al. Is variant angina the generalised manifestation of a generalised vasospastic disorder? N Engl J Med 1981; 304: 763-766.

70. Zamora MR, O'Brien RF, Rutherford RB, Weil JV. Serum endothelin-1 concentrations and cold provocation in primary Raynaud's phenomenon. Lancet 1990; 336: 1144-1147.

71. Kanno K, Hirata Y, Emori T et al. Endothelin and Raynaud's phenomenon. Am J Med 1991; 90: 130-131.

72. Toyo-oka T, Aizawa T, Suzuki N et al. Increased plasma level of endothelin-1 and coronary spasm induction in patients with vasospastic angina pectoris. Circulation 1991; 83: 476-483.

73. Knock GA, Terenghi G, Bunker CB et al. Characterization of endothelin-binding sites in human skin and their regulation in primary Raynaud's phenomenon and systemic sclerosis. J Invest Dermatol 1993; 101: 73-78.

74. Asano T, Ikegaki I, Suzuki Y et al. Endothelin and the production of cerebral vasospasm in dogs. Biochem Biophys Res Commun 1989; 159: 365-372.

75. Suzuki R, Masaoka H, Hirata Y et al. The role of endothelin-1 in the origin of cerebral vasospasm in patients with aneurysmal subarachnoid hemorrhage. J Neurosurg 1992; 77: 96-100.

76. Gaetani P, Rodriguez y Baena R, Grignani G et al. Endothelin and aneurysmal subarachnoid haemorrhage: a study of subarachnoid cisternal cerebrospinal fluid. J Neurol Neurosurg Psych 1994; 57: 66-72.

77. Clozel M, Breu V, Burri K et al. Pathophysiological role of endothelin revealed by the first orally active endothelin receptor antagonist. Nature 1993; 365: 759-761.

78. Foley PL, Caner HH, Kassell NF, Lee KS. Reversal of subarachnoid hemorrhage-induced vasoconstriction with an endothelin receptor antagonist. Neurosurgery 1994; 34: 108-113.

79. Watanabe T, Suzuki N, Shimamoto N et al. Endothelin in myocardial infarction. Nature 1990; 344: 114.

80. Watanabe T, Suzuki N, Shimamoto N et al. Contribution of endogenous endothelin to the extension of myocardial infarct size in rats. Circ Res 1991; 69: 370-377.

81. Liu J, Chen R, Casley DJ, Nayler WN. Ischaemia and reperfusion increase ^{125}I-labelled endothelin-1 binding in rat cardiac membranes. Am J Physiol 1990; 258: H829-H835.

82. McMurdo L, Thiemermann C, Vane JR. The effects of the endothelin ET_A receptor antagonist FR 139317, on infarct size in a rabbit model of acute myocardial ischaemia and reperfusion. Br J Pharmacol 1994; 113: 75-80.

83. Richard V, Knaeffer N, Hogie M et al. Role of endogenous endothelin in myocardial and coronary endothelial injury after ischaemia and reperfusion in rats: studies with bosentan, a mixed ET_A-ET_B antagonist. Br J Pharmacol 1994; 113: 869-876.

84. Haber E, Lee M-E. Endothelin to the rescue? Nature 1994; 370: 252-253.

85. Yasuda M, Kohno M, Tahara A et al. Circulating immunoreactive endothelin in ischaemic heart disease. Am Heart J 1990; 119: 801-806.

86. Stewart JT, Nisbet JA, Davies MJ. Plasma endothelin in coronary venous blood from patients with either stable or unstable angina. Br Heart J 1991; 66: 7-9.

87. Ray SG, McMurray JJ, Morton JJ, Dargie HJ. Circulating endothelin in acute ischaemic syndromes. Br Heart J 1992; 67: 383-386.

88. Weiczorek I, Haynes WG, Webb DJ et al. Plasma endothelin predicts outcome in patients with unstable myocardial ischaemic syndromes. Br Heart J 1994; 72: 436-441.

89. Miyauchi T, Yanagisawa M, Tomizawa T et al. Increased plasma concentrations of endothelin-1 and big endothelin-1 in acute myocardial infarction. Lancet 1989; 334: 53-54.

90. Weitzberg E, Lundberg JM, Rudehill A. Elevated plasma levels of endothelin in patients with sepsis syndrome. Circ Shock 1991; 33: 222-227.

91. Omland T, Lie RT, Aakvaag A et al. Plasma endothelin determination as a prognostic indicator of one year mortality after acute myocardial infarction. Circulation 1994; 89: 1573-1579.

92. Dargie HJ, McMurray JJV. Diagnosis and management of heart failure. Br Med J 1994; 308: 321-328.

93. The SOLVD Investigators. Effect of enalapril on survival in patients with reduced left ventricular ejection fractions and congestive heart failure. N Engl J Med 1991; 325: 293-302.

94. Cohn JN, Johnson G, Ziesche S et al. A comparison of enalapril with hydralazine-isosorbide dinitrate in the treatment of chronic congestive heart failure. N Engl J Med 1991; 325: 303-310.

95. Margulies KB, Hildebrand FL, Lerman A et al. Increased endothelin in experimental heart failure. Circulation 1990; 82: 2226-2230.

96. Teerlink JR, Löffler B-M, Hess P et al. Role of endothelin in the maintenance of blood pressure in conscious rats with chronic heart failure: acute

effects of the endothelin receptor antagonist Ro 47-0203 (Bosentan). Circulation 1994; 90: 2510-2518.

97. Hiroe M, Hirata Y, Fujita N et al. Plasma endothelin-1 levels in idiopathic dilated cardiomyopathy. Am J Cardiol 1991; 68: 1114-1115.

98. McMurray JJ, Ray SG, Abdullah I et al. Plasma endothelin in chronic heart failure. Circulation 1992; 85: 1374-1379.

99. Cody, RJ, Haas GJ, Binkley PF et al. Plasma endothelin correlates with the extent of pulmonary hypertension in patients with chronic congestive heart failure. Circulation 1992; 85: 504-509.

100. Pacher R, Bergler-Klein J, Globits S et al. Plasma big endothelin-1 concentrations in congestive heart failure patients with or without systemic hypertension. Am J Cardiol 1993; 71: 1293-1299.

101. Wei C-M, Lerman A, Rodeheffer RJ et al. Endothelin in human congestive heart failure. Circulation 1994; 89: 1580-1586.

102. Love MP, Haynes WG, Webb DJ, McMurray JJV. Endothelin converting enzyme (ECE) inhibition is of potential therapeutic benefit in chronic heart failure. Proceedings: International Meeting on Heart Failure-Therapeutic Targets 1994.

103. Roubert P, Gillard V, Plas P et al. Angiotensin II and phorbol-esters potently down-regulate endothelin (ET-1) binding sites in vascular smooth muscle cells. Biochem Biophys Res Commun 1989; 164: 809-815.

104. Kanno K, Hirata Y, Tsujino M et al. Up-regulation of ET_B receptor subtype mRNA by angiotensin II in rat cardiomyocytes. Biochem Biophys Res Commun 1993; 194: 1282-1287.

105. Haynes WG, Strachan FE, Webb DJ. Endothelin ET_A and ET_B receptors cause vasoconstriction of human resistance and capacitance vessels in vivo. Circulation 1995 (in press).

106. White DG, Garratt P, Mundin JW et al. (1994). Human saphenous vein contains both endothelin ET_A and ET_B receptors. Eur J Pharmacol 257: 307-310.

107. Seo B, Oemar BS, Siebenmann R et al. Both ET_A and ET_B receptors mediate contraction to endothelin-1 in human blood vessels. Circulation 1994; 89: 1203-1208.

108. Firth JD, Ratcliffe PJ, Raine AE., Ledingham JG. Endothelin: an important factor in acute renal failure? Lancet 1988; 2: 1179-1182.

109. Rakugi H, Tabuchi Y, Nakamaru M et al. Evidence for endothelin-1 release from resistance vessels of rats in response to hypoxia. Biochem Biophys Res Commun 1990; 169: 973-977.

110. Shibouta Y, Suzuki N, Shino A et al. Pathophysiological role of endothelin in acute renal failure. Life Sci 1990; 46: 1611-1618.

111. López-Farré A, Gómez-Garre D, Bernabeu F, López-Nova JM. A role for endothelin in the maintenance of post-ischemic acute renal failure in the rat. J Physiol (Lond) 1991; 444: 513-522.

112. Kon V, Yoshioka T, Fogo A, Ichikawa I. Glomerular actions of endothelin in vivo. J Clin Invest 1989; 83: 1762-1767.

113. Gellai M, Jugus M, Fletcher T et al. Reversal of post-ischemic acute renal failure with a selective endothelin$_A$ receptor antagonist in the rat. J Clin

Invest 1994; 93: 900-906.

114. Chan L, Chittinandana A, Shapiro JI et al. Effect of an endothelin-receptor antagonist on ischemic acute renal failure. Am J Physiol 1994; 266: F135-F138.

115. Tomita K, Ujiie K, Nakanishi T et al. Plasma endothelin levels in patients with acute renal failure. N Engl J Med 1989; 321: 1127-1127.

116. Moore K, Wendon J, Frazer M et al. Plasma endothelin immunoreactivity in liver disease and the hepatorenal syndrome. N Engl J Med 1992; 327: 1774-1778.

117. Petros A, Bennett D, Vallance P. Effect of nitric oxide synthase inhibitors on hypotension in patients with septic shock. Lancet 1991; 338: 1557-1558.

118. Kikeri D, Pennell JP, Hwang KH et al. Endotoxaemic acute renal failure in awake rats. Am J Physiol1986; 250: F1098-F1106.

119. Sugiura M, Inagami T, Kon V. Endotoxin stimulates endothelin-release in vivo and in vitro as determined by radioimmunoassay. Biochem Biophys Res Commun 1989; 161: 1220-1227.

120. Kon V, Badr KF. Biological actions and pathophysiologic significance of endothelin in the kidney. Kidney Int 1991; 40: 1-12.

121. Sanai L, Haynes WG, MacKenzie A et al. Endothelin production in sepsis and the adult respiratory distress syndrome. Crit Care Med 1995 (in press).

122. Brooks DP, Contino LC, Storer B, Ohlstein EH. Increased endothelin excretion in rats with renal failure induced by partial nephrectomy. Br J Pharmacol 1991; 104: 987-989.

123. Benigni A, Perico N, Gaspari F et al. Increased renal endothelin production in rats with reduced renal mass. Am J Physiol 1991; 260: F331-F339.

124. Orisio S, Benigni A, Bruzzi I et al. Renal endothelin gene expression is increased in remnant kidney and correlates with disease progression. Kidney Int 1993; 43: 354-358.

125. Nakamura T, Fukui M, Ebihara I et al. Effects of a low protein diet on glomerular endothelin family gene expression in experimental focal glomerular sclerosis. Clin Sci 1995; 88: 29-37.

126. Benigni A, Zoja C, Corna D et al. A specific endothelin subtype A receptor antagonist protects against injury in renal disease progression. Kidney Int 1993; 44: 440-445.

127. Koyama H, Tabata T, Nishzawa Y et al. Plasma endothelin levels in patients with uraemia. Lancet 1989; 333: 991-992.

128. Warrens AN, Cassidy MJD, Takahashi K et al. Endothelin in renal failure. Nephrol Dial Transplant 1990; 5: 418-422.

129. Ohta T, Hirata Y, Shichiri M et al. Urinary excretion of endothelin-1 in normal subjects and patients with renal disease. Kidney Int 1991; 39: 307-311.

130. Miyauchi T, Suzuki N, Kurihara T et al. Endothelin-1 and endothelin-3 play different roles in acute and chronic alterations of blood pressure in patients with chronic hemodialysis. Biochem Biophys Res Commun 1991; 178: 276-281.

131. Tsunoda K, Abe K, Yoshinaga K. Endothelin in hemodialysis-resistant hypertension. Nephron 1991; 59: 687-688.

132. Bunchman TE, Brookshire CA. Cyclosporine-induced synthesis of endothelin by cultured human endothelial cells. J Clin Invest 1991; 88: 310-314.

133. Kon V, Sugiura M, Inagami T et al. Role of endothelin in cyclosporine-induced glomerular dysfunction. Kidney Int 1990; 37: 1487-1491.

134. Nambi P, Pullen M, Contino LC, Brooks DP. Upregulation of renal endothelin receptors in rats with cyclosporin A-induced nephrotoxicity. Eur J Pharmacol 1990; 187: 113-116.

135. Takeda M, Breyer MD, Noland TD et al. Endothelin-1 receptor antagonist: effects on endothelin- and cyclosporin-treated mesangial cells. Kidney Int 1992; 42: 1713-1719.

136. Lanese DM, Conger JD. Effects of endothelin receptor antagonist on cyclosporin-induced vasoconstriction in isolated rat renal arterioles. J Clin Invest 1993; 91: 2144-2149.

137. Fogo A, Hellings SE, Inagami T, Kon V. Endothelin receptor antagonism is protective in in vivo acute cyclosporin toxicity. Kidney Int 1992; 42: 770-774.

138. Moutabarrik A, Ishibashi M, Fukunaga M et al. FK 506 mechanism of nephrotoxicity: stimulatory effect on endothelin secretion by cultured kidney cells and tubular cell toxicity in vitro. Tranplant Proc 1991; 23: 3133-3136.

INDEX

Page numbers in italics denote figures (f) or tables (t).

Molecular Biology
Intelligence Unit

Available and Upcoming Titles

NEUROSCIENCE
INTELLIGENCE UNIT

AVAILABLE AND UPCOMING TITLES

MEDICAL INTELLIGENCE UNIT
AVAILABLE AND UPCOMING TITLES

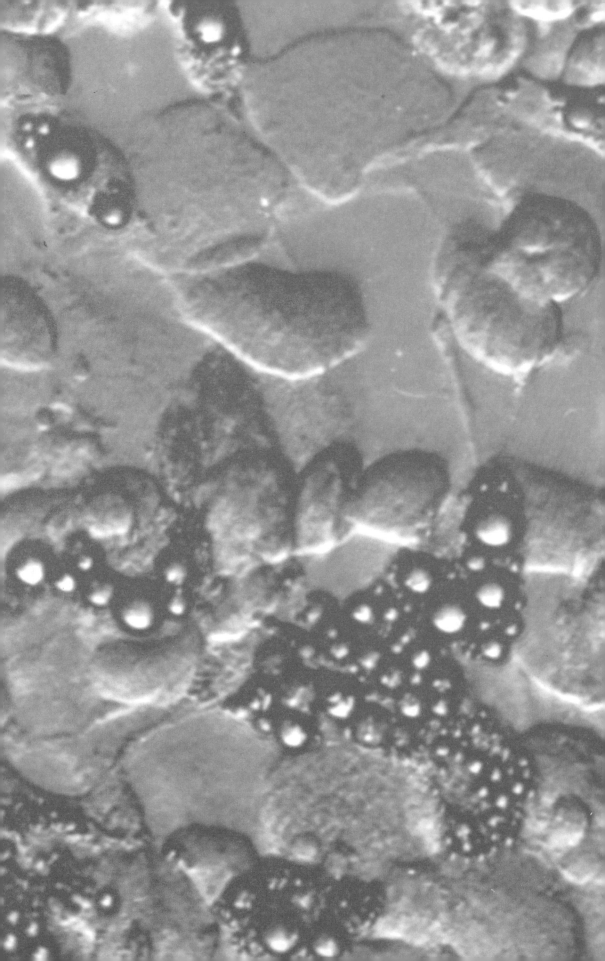